SpringerBriefs in Applied Sciences
and Technology

SpringerBriefs present concise summaries of cutting-edge research and practical applications across a wide spectrum of fields. Featuring compact volumes of 50 to 125 pages, the series covers a range of content from professional to academic.

Typical publications can be:

- A timely report of state-of-the art methods
- An introduction to or a manual for the application of mathematical or computer techniques
- A bridge between new research results, as published in journal articles
- A snapshot of a hot or emerging topic
- An in-depth case study
- A presentation of core concepts that students must understand in order to make independent contributions.

SpringerBriefs are characterized by fast, global electronic dissemination, standard publishing contracts, standardized manuscript preparation and formatting guidelines, and expedited production schedules.

On the one hand, **SpringerBriefs in Applied Sciences and Technology** are devoted to the publication of fundamentals and applications within the different classical engineering disciplines as well as in interdisciplinary fields that recently emerged between these areas. On the other hand, as the boundary separating fundamental research and applied technology is more and more dissolving, this series is particularly open to trans-disciplinary topics between fundamental science and engineering.

Indexed by EI-Compendex, SCOPUS and Springerlink.

Abdallah Mohamed Hamed

Image Processing Techniques for Deformed and Sparse Aperture Systems

 Springer

Abdallah Mohamed Hamed (ID)
Department of Physics
Faculty of Science
Ain Shams University
Cairo, Egypt

ISSN 2191-530X   ISSN 2191-5318   (electronic)
SpringerBriefs in Applied Sciences and Technology
ISBN 978-3-032-04920-9   ISBN 978-3-032-04921-6   (eBook)
https://doi.org/10.1007/978-3-032-04921-6

I dedicate this book to the spirit of my parents.

Preface

My recent publications on aperture modulation and its applications to the confocal laser microscope and speckle imaging have led to the presentation of this new book, *Image Processing Techniques for Deformed and Sparse Aperture Systems*. We intended this book for graduate students in optical sciences and optical engineering.

The object of drafting a book on image processing for deformed and sparse aperture systems is outlined below.

The book is composed of seven chapters. Chapter 1 discusses modulated sparse aperture imaging systems (MSAIS) using linear and quadratic apertures, as well as their applications to speckle images. Chapter 2 deals with hexagonal sparse apertures surrounded by a rectangular annulus.

A heterogeneous aperture and cascaded black and quadratic distributions are presented in Chaps. 3 and 4. Chapter 5 discusses the design of cascaded linear-conic apertures and hexagonal apertures, as well as their applications in microscopy. In addition, the design of a STRAUBEL filter in a circular aperture and its application on a confocal laser scanning microscope (CLSM) are presented in Chap. 6.

Finally, Chap. 7 investigates modulated cardiac apertures and their applications in confocal laser scanning microscopes.

Cairo, Egypt   Abdallah Mohamed Hamed

Competing Interests The author has no competing interests to declare that are relevant to the content of this manuscript.

Contents

Contents

Chapter 1
Investigation of Modulated Sparse Aperture Imaging Systems (MSAIS) and Their Applications on Speckle Images

1.1 Introduction

Optical sparse aperture (OSA) imaging systems can produce high-resolution images while maintaining a light-collecting area smaller than a filled aperture, achieving the same resolution [1–4].

Many optimization methods have been proposed in previous research. Stokes and Duncan improved the mid-frequency MTF by changing the radius of sub-apertures in the classical aperture structure [5]. Different algorithms have been proposed to optimize the MTF, including a genetic algorithm [3] and a minimum image reconstruction error criterion [6]. Other algorithms optimize the position and size of the sub-aperture by examining the distribution of the spatial frequency spectrum in a particular frequency domain [7].

A high-speed and efficient image restoration method for synthetic aperture systems is obtained using the deconvolution [8, 9]. Recently, a non-blind deconvolution algorithm for image restoration of optical synthetic aperture systems has been proposed in [10]. Other methods for resolution improvement in a microscope using amplitude modulation are outlined in [11–15].

Recently, the authors in [16] demonstrated an approach for image restoration in an actual multi-aperture system. A binocular telescope testbed consisting of two horizontally arranged collector telescopes with a 127 mm diameter has been built in their laboratory. Synthetic images using different resolution test charts with horizontal resolution improved were acquired after the two sub-apertures were co-phased.

In this chapter, we calculate the lateral point spread function for the sparse aperture imaging system (SAIS). We considered five sub-apertures surrounded by an annulus, and we deduced the total bandwidth (BW) for the PSF instead of the investigation of the modulation transfer function (MTF) realized in the previous work [5]. In addition, we prove that the BW is decreased for the full-filled sub-apertures inside the circle; hence we have an improvement in the microscope's resolution. We computed the contrast of the speckle images corresponding to the modulated sparse apertures, considering seven sub-apertures.

1.2 Analysis

1.2.1 Computation of the PSF for the Sparse Apertures

The pupil function of a sparse aperture imaging system (SAIS) modulated by an annulus is represented as follows:

$$P_T(x, y) = P_1(x, y) + P_2(x - x_0, y) + P_3\left(x - \frac{x_0}{2}, y - y_0\right)$$

$$+ P_4\left(x + \frac{3x_0}{2}, y - y_0\right) + P_5\left(x + \frac{3x_0}{2}, y + y_0\right) + \text{annul}(x, y) \quad (1.1)$$

We assumed five equal sub-apertures $P_1, P_2, \ldots, P_5$ surrounded by an annulus of external radius $= \rho_0$ and internal radius $= \rho_0 - \delta\rho$. The SAIS aperture is like the Golay-5 aperture.

The modulated SAIS aperture, adding the annulus in Eq. (1.1), is rewritten as

$$P_T(x, y) = P_{sub}(x, y) \otimes \left[\delta(x, y) + \delta(x - x_0, y) + \delta\left(x - \frac{x_0}{2}, y - y_0\right)\right.$$

$$\left. + \delta\left(x + \frac{3x_0}{2}, y - y_0\right) + \delta\left(x + \frac{3x_0}{2}, y + y_0\right)\right] + \left[P_{ext}(x, y) - P_{int}(x, y)\right] \quad (1.2)$$

where $P_1 = P_2 = \cdots = P_5 = P_{sub}$

We computed the point spread function (PSF) using the Fourier transform on Eq. (1.2). We obtained the PSF as follows:

$$h(u, v) = F.T. \left\{P_{sub}(x, y) \otimes \left[\delta(x, y) + \delta(x - x_0, y)\right.\right.$$

$$\left.\left. + \delta\left(x - \frac{x_0}{2}, y - y_0\right) + \delta\left(x + \frac{3x_0}{2}, y - y_0\right) + \delta\left(x + \frac{3x_0}{2}, y + y_0\right)\right]\right\} \quad (1.3)$$

We make use of the properties of the Fourier transform and convolution operation; hence, Eq. (1.3) becomes as follows:

$$h(u, v) = F.T. \left[P_{sub}(x, y) \right] . F.T. \left[\delta(x, y) + \delta(x - x_0, y) + \delta\left(x - \frac{x_0}{2}, y - y_0\right) \right.$$

$$\left. + \delta\left(x + \frac{3x_0}{2}, y - y_0\right) + \delta\left(x + \frac{3x_0}{2}, y + y_0\right) \right]$$

$$+ \left[\frac{J_1(w)}{w} - \epsilon^2 \frac{J_1(\epsilon w)}{\epsilon w} \right] \tag{1.4}$$

In Eq. (1.4), $\varepsilon = 1 - \frac{\delta\rho}{\rho_0}$ represents the ratio of the internal to the external radius for the uniform circle of radius ρ_0. The Fourier transform of the sub-aperture of a radius $= \rho_s = \frac{\rho_0}{16} = 16$ pixels for the radius of the whole circle is 256 pixels.

Finally, we performed the transformation in the shifted Dirac delta function in Eq. (1.4) to obtain the PSF as follows:

$$h(u, v) = \left[\frac{J_1(w)}{w} - \epsilon^2 \frac{J_1(\epsilon w)}{\epsilon w} \right] + \left[\frac{J_1\left(\frac{w}{16}\right)}{\frac{w}{16}} \right] .$$

$$\left\{ 1 + \exp\left(-\frac{j2\pi}{\lambda f} x_0 u\right) + \exp\left[-\frac{j2\pi}{\lambda f}\left(\frac{x_0}{2}u + y_0 v\right)\right] \right.$$

$$\left. + \exp\left[\frac{j2\pi}{\lambda f}\left(\frac{3x_0}{2}u - y_0 v\right)\right] + \exp\left[\frac{j2\pi}{\lambda f}\left(\frac{3x_0}{2}u + y_0 v\right)\right] \right\} \tag{1.5}$$

In Eq. (1.5), we assumed that $\frac{2\pi}{\lambda f} x_0 = \alpha$ and $\frac{2\pi}{\lambda f} y_0 = \beta$. Then, the real part of the PSF is written as follows:

$$h_{real}(u, v) = \left[\frac{J_1(w)}{w} - \epsilon^2 \frac{J_1(\epsilon w)}{\epsilon w} \right] + \left[\frac{J_1\left(\frac{w}{16}\right)}{\frac{w}{16}} \right]$$

$$\left\{ 1 + \cos(\alpha u) + \cos\left(\frac{\alpha}{2}u + v\right) + \cos\left(\frac{3\alpha}{2}u - \beta v\right) + \cos\left(\frac{3\alpha}{2}u + \beta v\right) \right\} \tag{1.6}$$

The imaginary part of the amplitude PSF is written as follows:

$$h_{imag}(u, v) = \left[\frac{J_1(w)}{w} - \epsilon^2 \frac{J_1(\epsilon w)}{w} \right] . \left[\frac{J_1\left(\frac{w}{16}\right)}{\frac{w}{16}} \right]$$

$$\left\{ -\sin(\alpha u) - \sin\left(\frac{\alpha}{2}u + v\right) + \sin\left(\frac{3\alpha}{2}u - \beta v\right) + \sin\left(\frac{3\alpha}{2}u + \beta v\right) \right\} \tag{1.7}$$

The corresponding intensity $I(u, v)$ is the modulus square of the amplitude PSF given by the following:

$$I(u, v) = \mid h_{real}(u, v) + j\, h_{imag}(u, v) \mid^2 = \mid h_{real}(u, v) \mid^2 + \mid h_{imag}(u, v) \mid^2 . \tag{1.8}$$

i. For the linear distribution of sub-apertures, we obtained the PSF as follows:

$$h_{linear}(w) = F.T.\{\rho\} = \left[\frac{J_1\left(\frac{w}{16}\right)}{\frac{w}{16}}\right] + \left[\frac{J_0\left(\frac{w}{16}\right)}{\left(\frac{w}{16}\right)^2}\right] - 2\left[\frac{\sum_i J_i\left(\frac{w}{16}\right)}{\left(\frac{w}{16}\right)^2}\right] \tag{1.9}$$

Hence, $h_{linea}(w)$ replaces the PSF for the uniform circular sub-aperture in Eq. (1.5).

ii. For the quadratic distribution of sub-apertures, we obtained the PSF as follows:

$$h_{quadratic}(w) = F.T.\{\rho^2\} = \left[\frac{J_1\left(\frac{w}{16}\right)}{\left(\frac{w}{16}\right)}\right] - 2\left[\frac{J_2\left(\frac{w}{16}\right)}{\left(\frac{w}{16}\right)^2}\right] \tag{1.10}$$

In this case, $h_{quadrati}(w)$ replaces the PSF in equation (1.5).

iii. For two different distributions, e.g., one is a uniform circular aperture and the other has a quadratic distribution, we obtained the PSF depending on its distribution and position in the sparse aperture.

1.2.2 Application of Speckle Imaging and Computation of the Image Contrast

The sparse aperture in Eq. (1.1) is illuminated with a coherent laser beam followed by a diffuser d (x, y). The diffuser is represented in matrix form as follows:

$$d(x, y) = \sum_{m=1}^{M} \sum_{n=1}^{N} \text{rand}(m\Delta x, n\Delta y) \tag{1.11}$$

Hence, we represent the product of the amplitude transmittance and the diffuser as follows:

$$A(x, y) = P_T(x, y).d(x, y) \tag{1.12}$$

We obtained the speckle pattern in the focal plane of a converging lens by performing the FFT upon Eq. (1.12), as follows:

$$B(u, v) = F.T.\{A(x, y)\} = F.T.\{P_T(x, y).\, d(x, y)\} = h(u, v) \otimes \tilde{d}(u, v) \tag{1.13}$$

$$\otimes : \text{symbol for convolution.}$$

$\tilde{d}(u, v) = F.T.\{d(x, y)\}$, and $h(u, v) = F.T.\{P_T(x, y)\}$

Hence, the point spread function represented by Eq. (1.5) affects the speckle pattern represented by Eq. (1.13).

Since the intensity is the modulus square of the Eq. (1.13), the speckle image is represented as follows:

$$I(u, v) = \mid h(u, v) \otimes \tilde{d}\,(u, v) \mid^{2} \tag{1.14}$$

The image contrast corresponding to the speckle pattern is computed using the following formula:

$$K = \frac{\sigma}{<I>} \tag{1.15}$$

σ is the root mean square deviation from the mean value $<I>$.

1.3 Results and Discussion

The design of a sparse aperture of five sub-circular apertures with a linear distribution is shown in Fig. 1a. The corresponding image of a point array autocorrelation with 21 points is shown in Fig. 1b. The autocorrelation plot at $x = 512$ pixels is shown in Fig. 1c, and the autocorrelation plot at $y = 512$ pixels is shown in Fig. 1d. The bandwidth equals twice the diameter corresponding to the sub-apertures of $2 \times 64 = 128$ pixels. Referring to the autocorrelation image in (b), we obtained two peaks beside the central peak at a constant x of 512 pixels, shown in (c). We obtained four peaks around the central peak at a constant y of 512 pixels, shown in (d).

Four peaks are obtained as in Fig. 1.2. We showed that the number of autocorrelation peaks depends on the measurement direction. For example, we have only two peaks at constant x, while we have four peaks at y equal to a constant, as shown in Fig. 1c and d. The distribution of the autocorrelation peak is affected by the distribution of the sub-apertures, either linear or quadratic.

In both autocorrelation plots, the bandwidth is twice the diameter corresponding to the sub-apertures of $2 \times 64 = 128$ pixels.

We applied the FFT technique to compute the PSF. An image of the lateral PSF corresponding to the sparse aperture with linear distribution for the sub-apertures is presented in Fig. 3a and b. In (c), the lateral PSF plot is at a constant $y = 512$ pixels. In (d), the PSF plot is shown at x of 512 pixels.

Another result for the lateral PSF corresponding to the quadratic distribution of the sub-apertures in the sparse aperture is shown in Fig. 1.4. We obtained similar results for the lateral PSF. We have the same cut-off spatial frequency in both cases of linear and quadratic distributions. The results in Figs. 1.1, 1.2, 1.3, 1.4 concerning the sparse aperture assume only the cases of linear and quadratic sub-apertures in the absence of the annulus surrounding the sub-apertures.

The normalized lateral PSF plots corresponding to the Golay-5 sub-apertures of radii $= 16, 32, 40, 48, 56,$ and 64 pixels are shown in Fig. 5a–f. All plots are taken from $x = 960$ to $x = 1088$ pixels at $y = 1024$ pixels. The total width of the central peak $= 13$ pixels (Table 1.1).

Figure 1.6a–c shows the normalized lateral PSF plots. We used Eqs. (1.6) and (1.7), corresponding to the Golay-5 sub-apertures, for a radius of 16 pixels. The

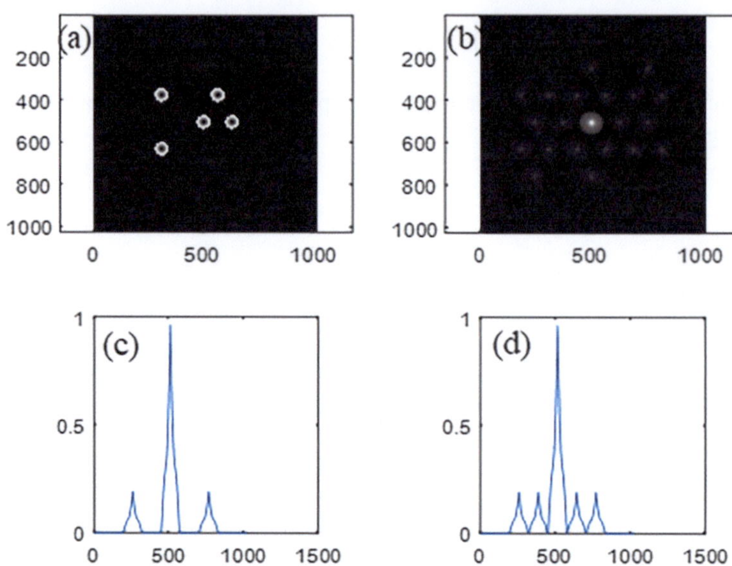

Fig. 1.1 In **a** the sparse aperture of five sub-linear apertures (Golay-5). In **b** the point array auto-correlation of 21 points. In **c** the autocorrelation plot at x = 512 pixels, and in **d** the autocorrelation plot at y = 512 pixels. The bandwidth equals twice the diameter corresponding to the sub-apertures of $2 \times 64 = 128$ pixels

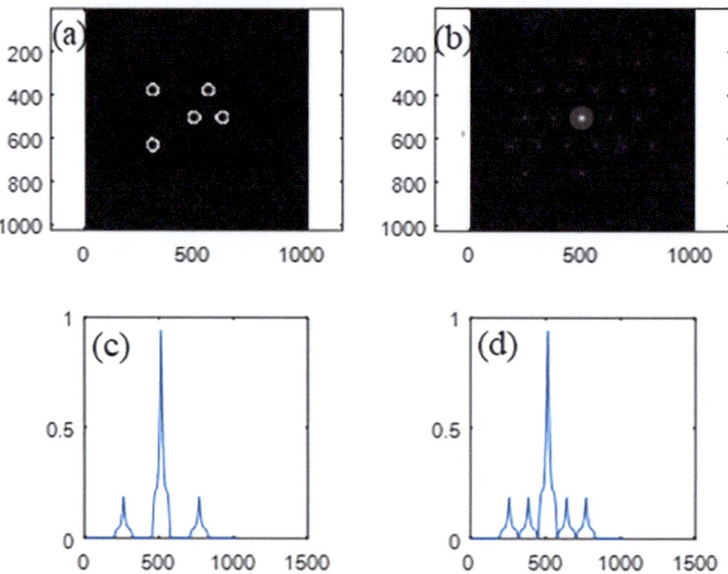

Fig. 1.2 In **a** the sparse aperture of five sub-quadratic apertures (Golay-5). In **b** the point array auto-correlation of 21 points. In **c** the autocorrelation plot at x = 512 pixels, and in **d** the autocorrelation plot at y = 512 pixels

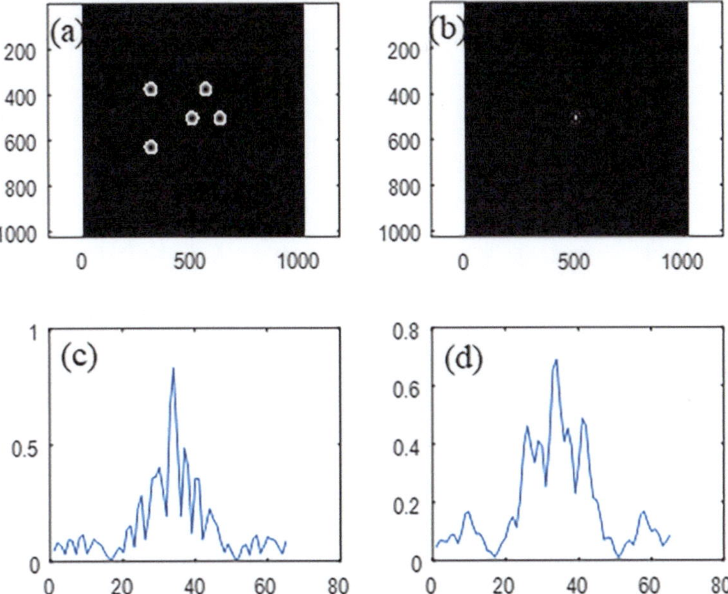

Fig. 1.3 In **a** the sparse aperture of five sub-apertures of linear distribution. In **b** the lateral PSF image corresponds to the sub-apertures of the linear distribution. In **c** the PSF plot is at constant y = 512 pixels, and in **d** the plot is at constant x = 512 pixels

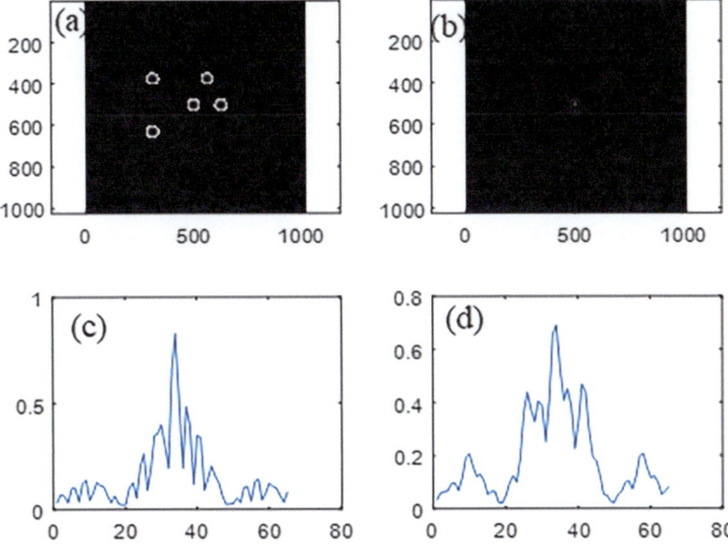

Fig. 1.4 In **a** the sparse aperture of five sub-apertures of quadratic distribution. In **b** the lateral PSF image corresponds to the sparse aperture shown in **a**. In **c** the PSF plot is at constant y = 512 pixels, and in **d** the PSF plot is at constant x = 512 pixels

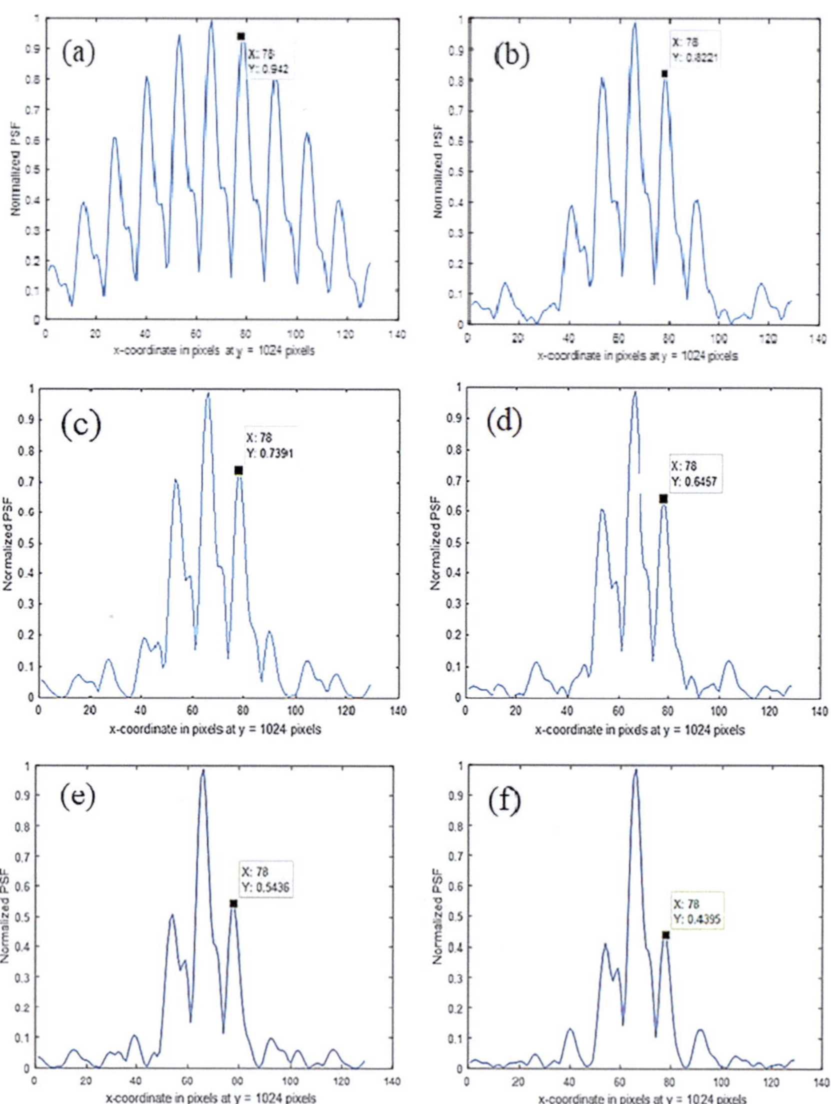

Fig. 1.5 The normalized lateral PSF plots, computed using the FFT method, corresponding to the Golay-5 sub-apertures of the radius = 16, 32, 40, 48, 56, and 64 pixels shown in Fig. 1.5a–f. All plots are taken from x = 960 to x = 1088 pixels at y = 1024 pixels. The total width of the central peak is = 13 pixels

computed bandwidth of the central peak is $=15$ pixels. Figure 1.6a and b shows the real and imaginary plots for the amplitude PSF. All theoretical plots in Fig. 1.6 are taken from $x = -56$ to $x = 256$ pixels.

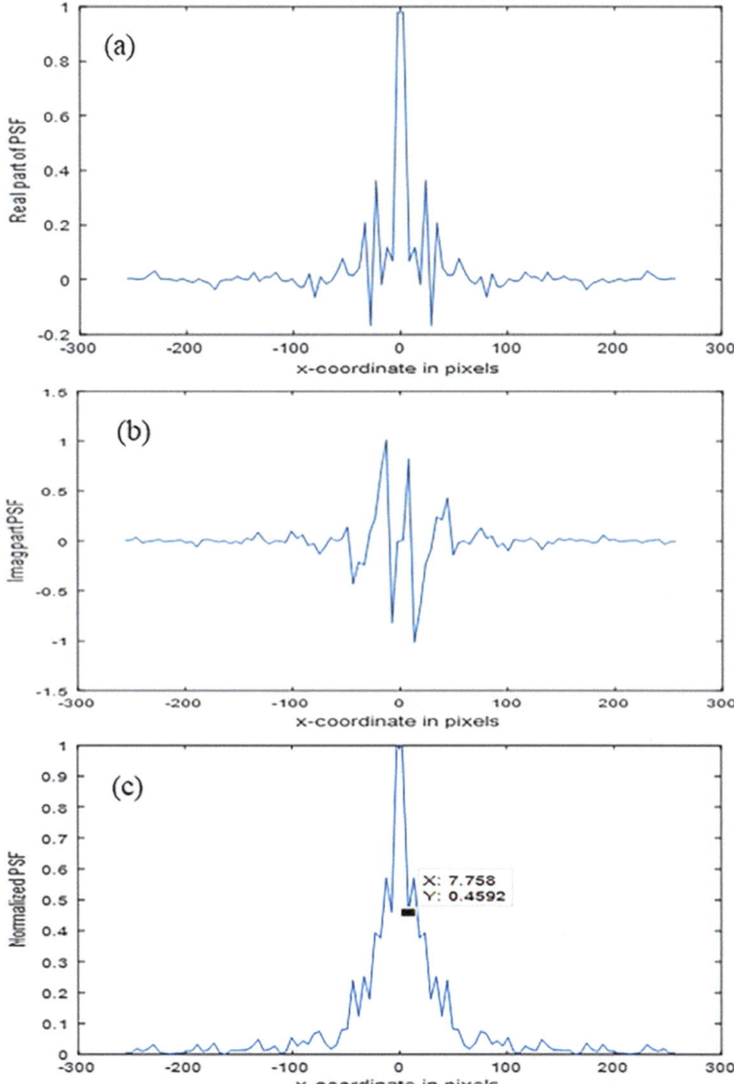

Fig. 1.6 The normalized lateral PSF plots computed from Eqs. (1.6) and (1.7), corresponding to the Golay-5 sub-apertures of radius $= 16$ pixels shown in Fig. 9.6c. All plots are taken from $x = -256$ to $x = 256$ pixels. The computed bandwidth of the central peak is $= 15$ pixels. In Fig. 1.6a, b. The real and imaginary plots for the amplitude PSF are shown

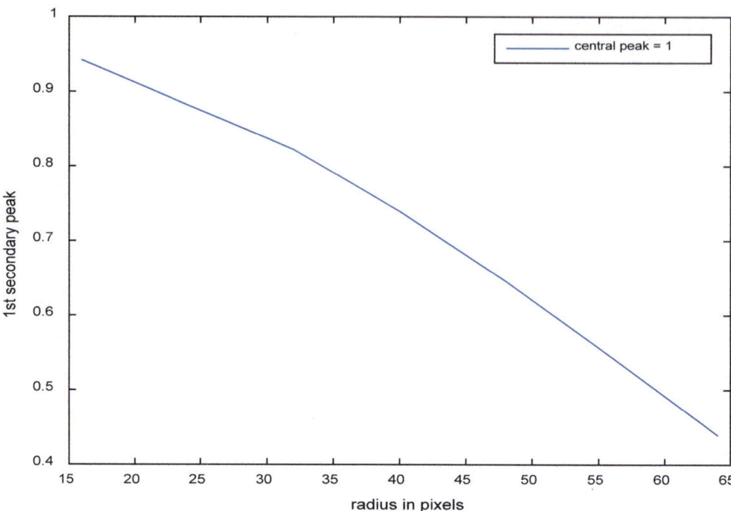

Fig. 1.7 A plot of the first secondary peak versus the radius of the sub-apertures in pixels in the range from 16 pixels up to 64 pixels

A plot of the first secondary peak versus the radius of the sub-apertures in pixels in the range from 16 to 64 pixels is shown in Fig. 1.7. It is shown that the height of the first secondary peak decreases with the increase in the radius of the sub-apertures. However, the Golay-5 sub-apertures are useful in imaging microscopic extended objects for the increased height of the secondary peaks attained at a small value of the radius of the sub-apertures.

We present the lateral PSF and the autocorrelation image where an annulus surrounds the sub-apertures.

The lateral PSF pattern corresponds to a sparse aperture surrounded by an annulus with a width of 16 pixels, as shown in Fig. 1.8a and b. In the second row, the lateral PSF plots at $x = 512$ pixels and $y = 512$ pixels are shown in Fig. 1.8c, d. We have legs that are affected by the presence of the annulus. It is compared to the results obtained in the absence of the aperture as in Figs. 1.4, 1.5, 1.6. The legs are strengthened as in Fig. 1.8c and d. The strengthened legs are useful for imaging extended objects in conventional and confocal microscopes. The sparse sub-apertures have a radius of 32 pixels.

The point array autocorrelation (PAA) image corresponds to the sparse aperture of five sub-apertures surrounded by an annulus, and is represented in Fig. 1.9a and b. The PAA plots at constant $y = 512$ pixels and constant $x = 512$ pixels are shown in Fig. 1.9c and d.

We used the FFT method to compute the lateral PSF corresponding to a sparse aperture composed of seven sub-apertures, each with a radius $= 16$ pixels. We plotted the normalized PSF at constant $y = 512$ pixels and computed the total $BW = 7$ pixels as shown in Fig. 1.10a. Two sub-apertures are placed outside of the circle of radius

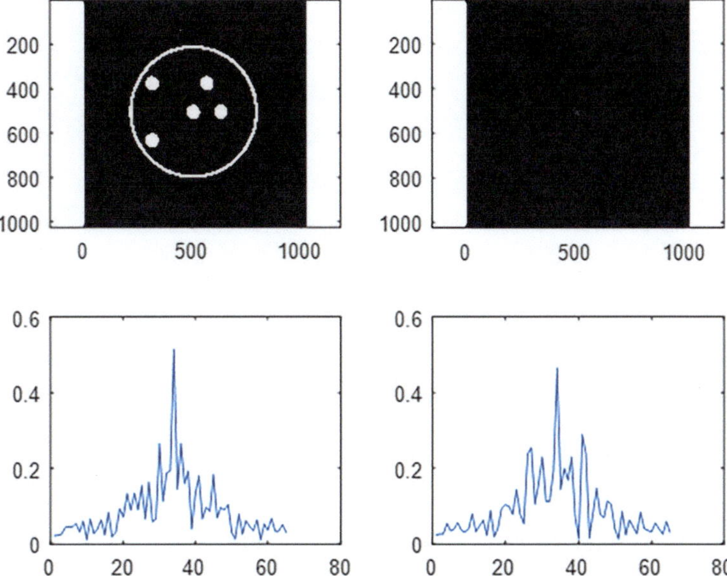

Fig. 1.8 In the first row, the lateral PSF pattern corresponds to the sparse aperture surrounded by an annulus of width = 16 pixels. In the second row, the PSF plots at x = 512 pixels and y = 512 pixels. The radius of the sub-apertures = 32 pixels. The external radius of the whole aperture is 300 pixels

= 256 pixels. When we displaced the two sub-apertures within the circle of the same radius, we obtained a sharper peak with total BW = 4 pixels, as shown in Fig. 1.10b. Hence, we have an improved PSF when all sub-apertures are filled with the circular aperture. Then, we have an improved resolution for the microscope, either conventional or confocal.

Design of a sparse aperture composed of 7 sub-apertures of two different distributions, where three uniform circular and four linear distributions are shown in Fig. 1.11. Each of the sub-apertures has a radius = 64 pixels. The matrix has dimensions 2048 × 2048 pixels.

The logarithmic scale of the impulse response corresponding to the aperture shown in Fig. 1.11 is represented in Fig. 1.12. The matrix has dimensions 512 × 512 pixels. The line plot of the normalized PSF corresponding to the sparse aperture at y = 256 pixels is shown in Fig. 1.13. The x-coordinate ranges from 224 to 288 pixels around the center at 256 pixels. We have the secondary peaks that are sharp and intense compared to the central peak.

We plotted the logarithmic scale of the impulse response corresponding to the aperture of Fig. 1.14 in Fig. 1.15. The matrix has dimensions 512 × 512 pixels. We draw a line plot of the normalized PSF corresponding to the sparse aperture at y = 256 pixels as in Fig. 1.16. The x-values were extended from 224 to 288 pixels and centered at 256 pixels.

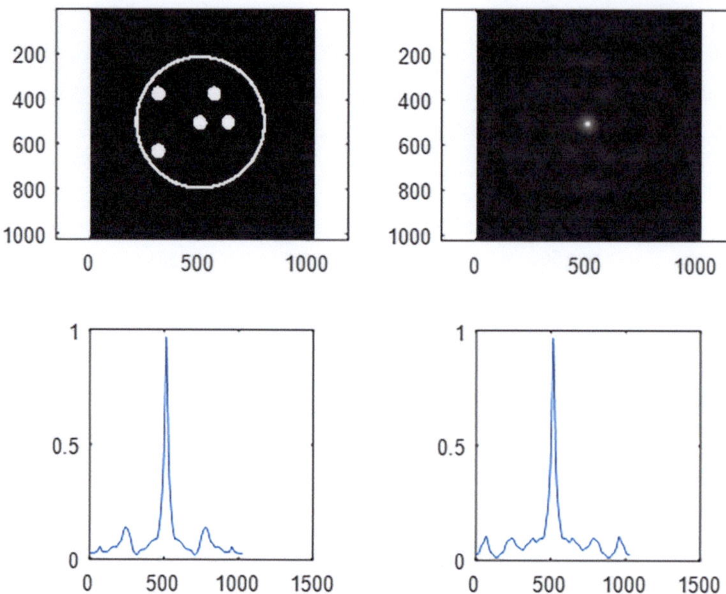

Fig. 1.9 In **a** the sparse aperture of five sub-apertures is surrounded by an annulus. In **b** the PAA image corresponds to the aperture. In **c** the PAA plot at y = 512 pixels, and in **d** the PAA plot at x = 512 pixels

We have, as in Fig. 1.17a, that the autocorrelation image corresponding to the sparse aperture in Fig. 1.11 was computed using the FFT. We plotted the autocorrelation image in Fig. 1.17b. It is obtained from the direct autocorrelation of the sparse aperture shown in Fig. 1.11. The three-dimensional autocorrelation intensity corresponding to the sparse aperture in Fig. 1.11 is plotted in Fig. 1.18. The values are resized to 0.2, 0.3, and 0.4, respectively.

In Fig. 1.19, the autocorrelation image corresponding to the direct autocorrelation of the sparse aperture is shown in Fig. 1.14.

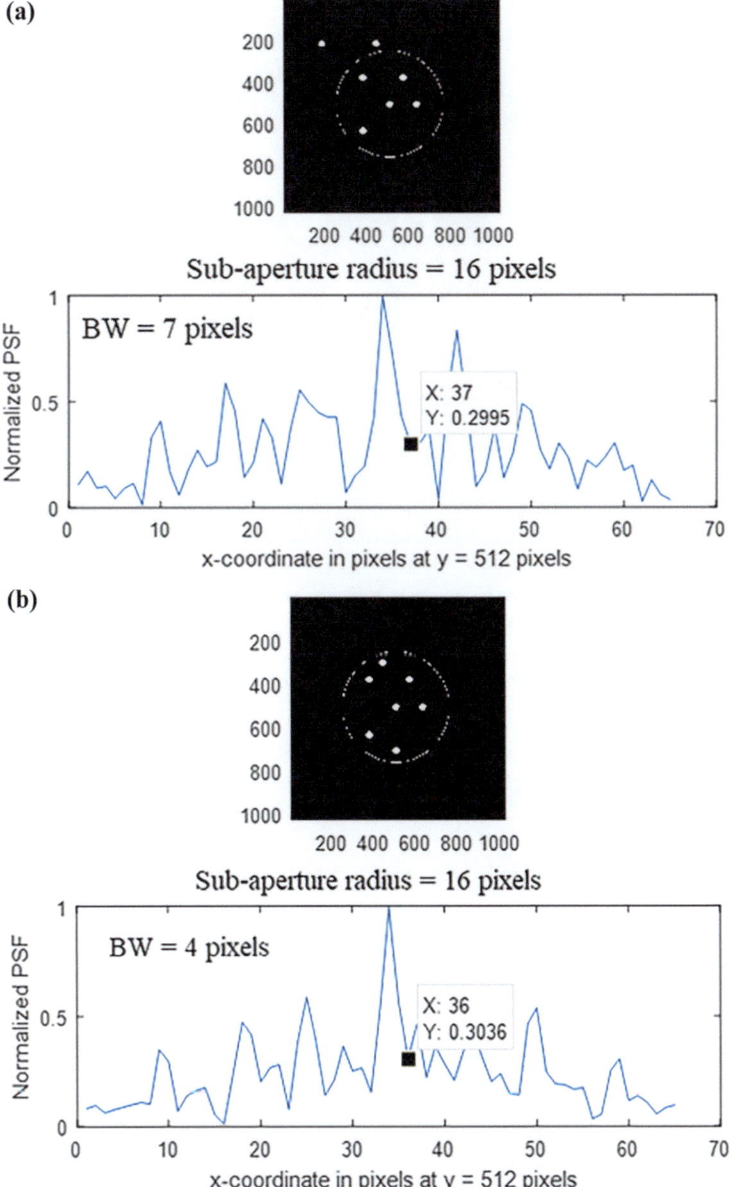

Fig. 1.10   a The normalized PSF versus x-coordinate at y = 512 pixels. The computed central peak of total BW from the graph is equal to 37−30 = 7 pixels. The aperture has seven sub-apertures. Five of them are inside the circle of radius 512 pixels, and the other two are outside the circle, as shown in the figure. Sub-aperture radius = 16 pixels **b** The normalized PSF versus x-coordinate at y = 512 pixels. The graph's computed central peak of total BW is = 36−32 = 4 pixels. The aperture has seven sub-apertures inside the circle of radius = 512 pixels as shown in the figure

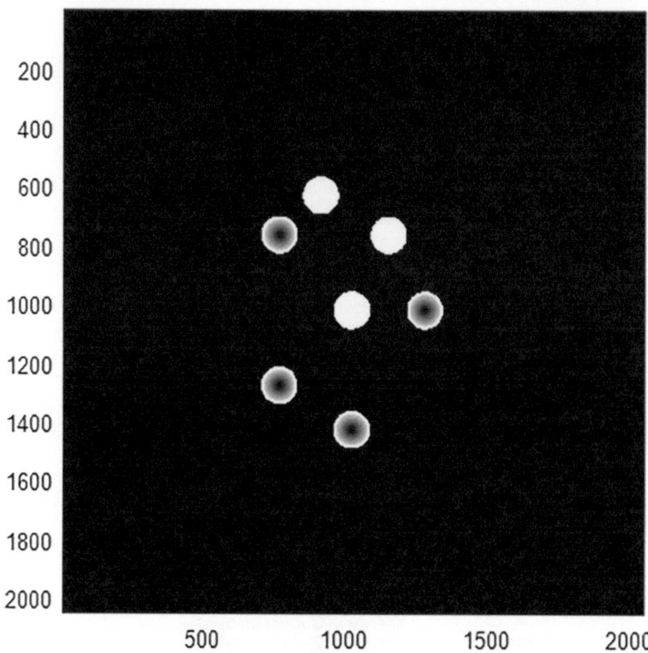

Fig. 1.11 A sparse aperture composed of 7 sub-apertures of two different distributions, with three uniform circular and four linear distributions. Each of the sub-apertures has a radius = 64 pixels. The matrix has dimensions 2048 × 2048 pixels

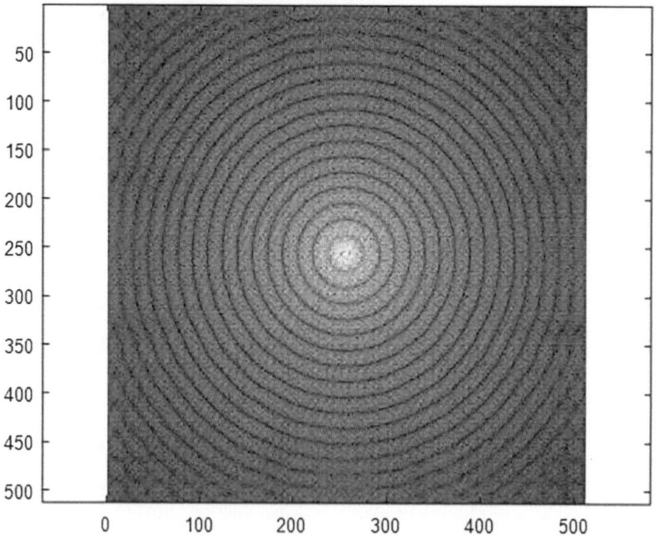

Fig. 1.12 Logarithmic scale of the impulse response corresponding to the aperture shown in Fig. 1.11. The matrix has dimensions 512 × 512 pixels

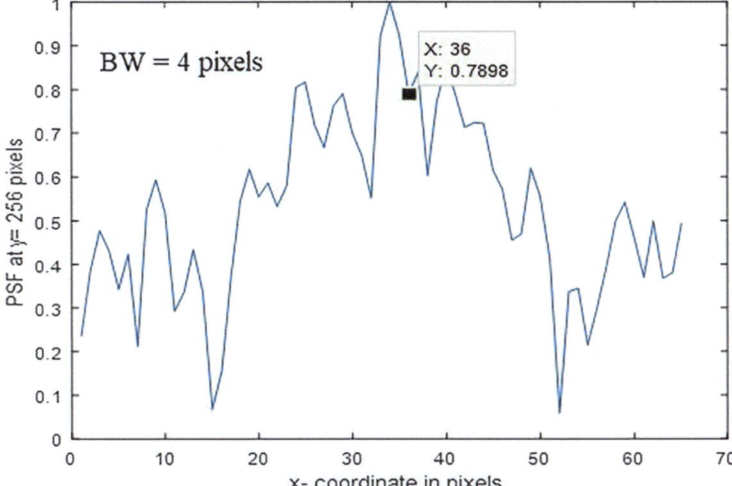

Fig. 1.13 Line plot of the normalized PSF corresponding to the sparse aperture at $y = 256$ pixels. The x-coordinate ranges from 224 to 288 pixels around the center at 256 pixels

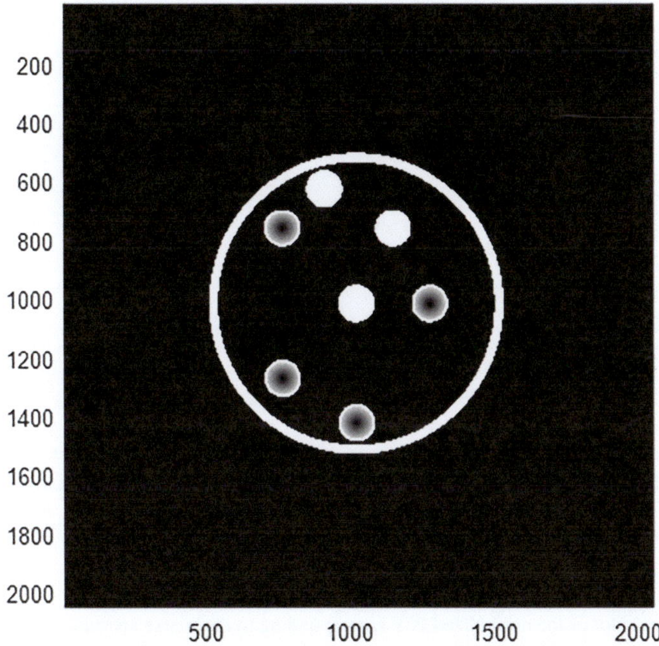

Fig. 1.14 The aperture in Fig. 1.14 is surrounded by an annulus of width $= 32$ pixels, and the external radius $= 512$ pixels. Each of the sub-apertures has a radius $= 64$ pixels. The total matrix has dimensions equal to 2048×2048 pixels

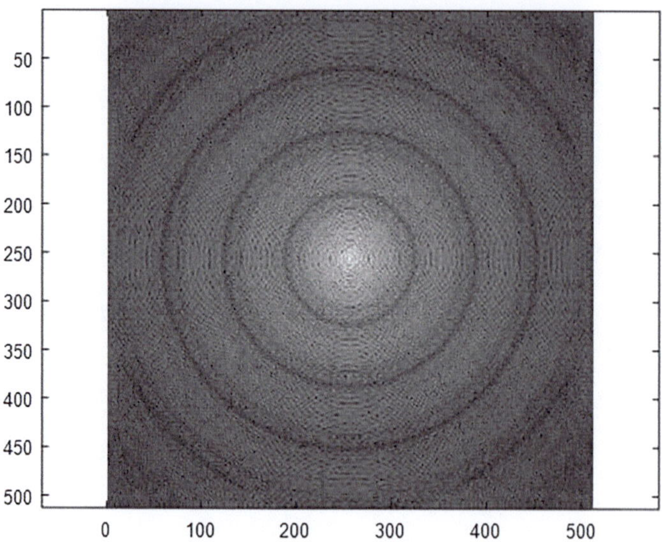

Fig. 1.15 Logarithmic scale of the impulse response corresponding to the aperture shown in Fig. 1.14. The whole matrix has dimensions 512×512 pixels

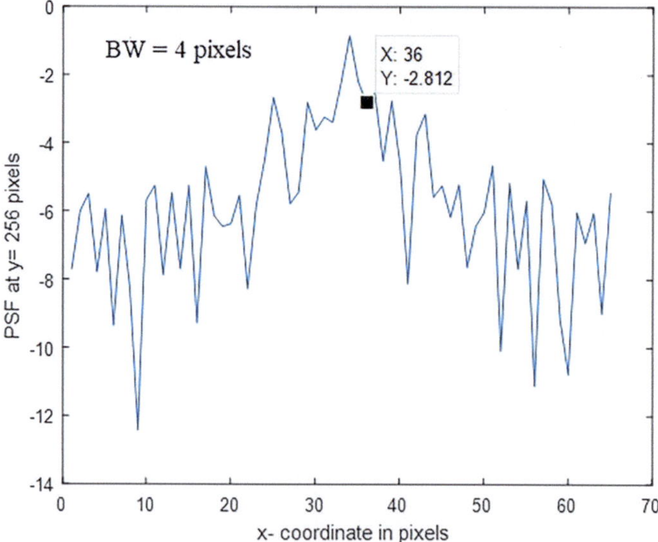

Fig. 1.16 Line plot of the normalized PSF corresponding to the sparse aperture at $y = 256$ pixels. The x-range is 224 to 288 pixels. The center is at 256 pixels

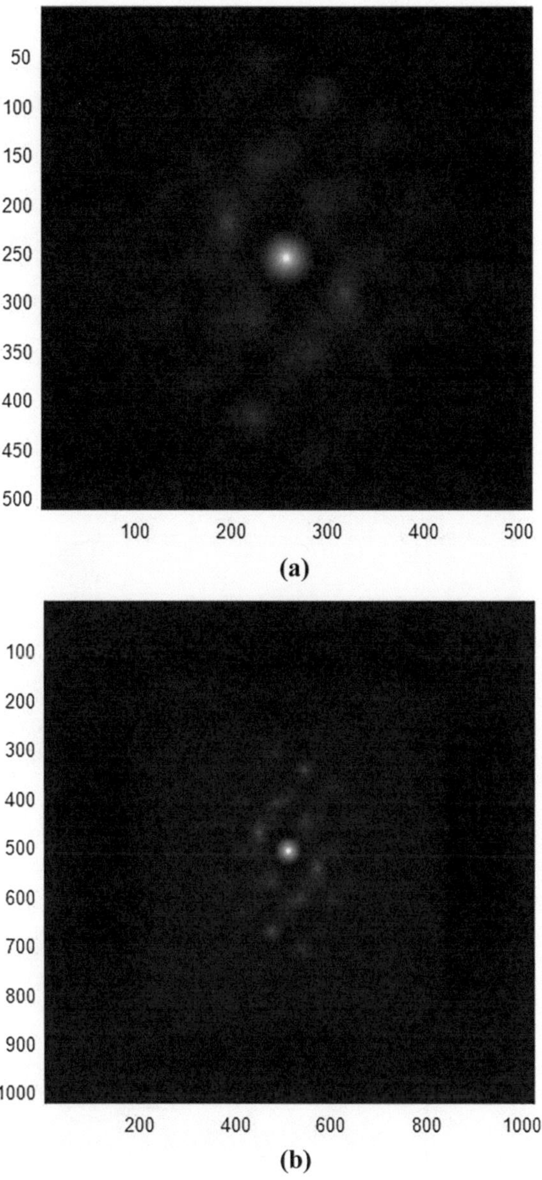

Fig. 1.17 **a** Autocorrelation image corresponding to the sparse aperture computed using FFT **b** Autocorrelation image corresponding to the direct autocorrelation of the sparse aperture shown in Fig. 1.11

The speckle image is obtained by performing the FFT on the diffuser, followed by a sparse aperture. The speckle images corresponding to the sparse apertures with seven sub-apertures each has a radius of 16 pixels. Two sub-apertures outside the circle of radius $= 256$ pixels are shown in Fig. 1.20. The speckle image in Fig. 1.20c corresponds to the diffraction of the sparse aperture without the annulus in Fig. 1.20a. The speckle image shown in Fig. 1.20d corresponds to the diffraction of the sparse

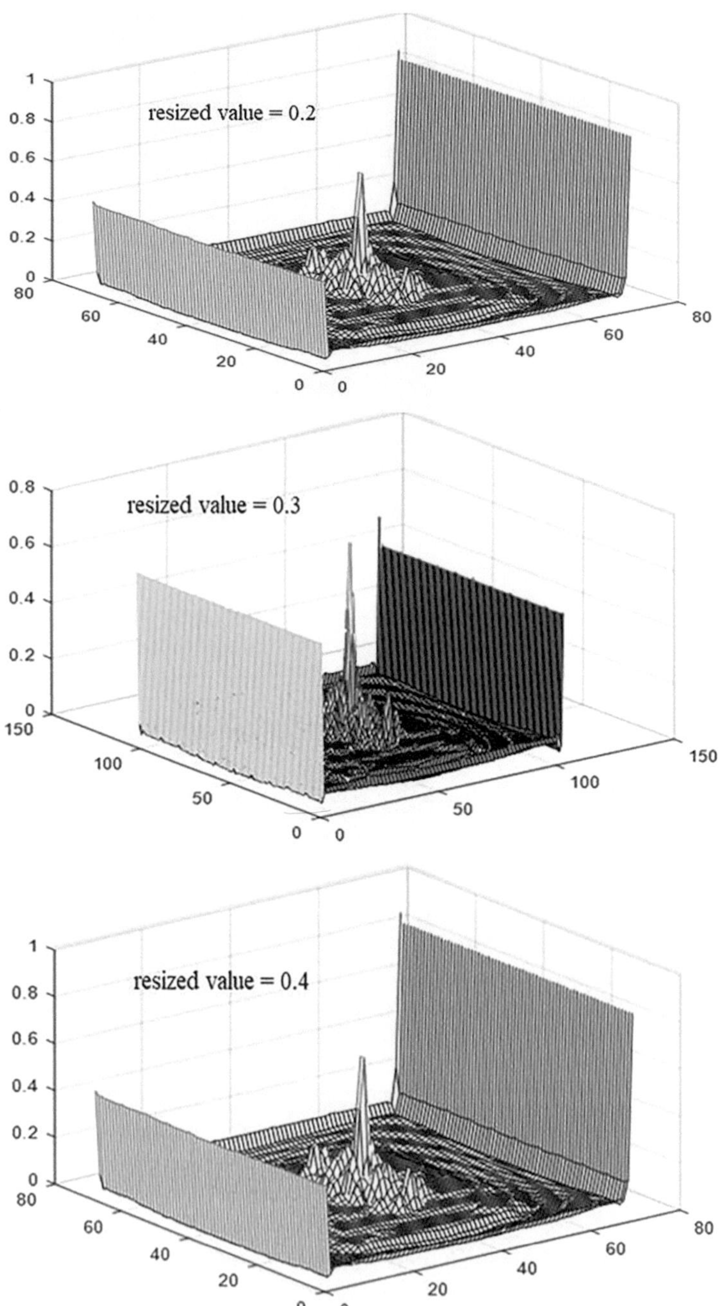

Fig. 1.18 Three-dimensional plot of the autocorrelation intensity corresponding to the sparse aperture shown in Fig. 1.11 The values are resized to 0.2, 0.3, and 0.4, respectively

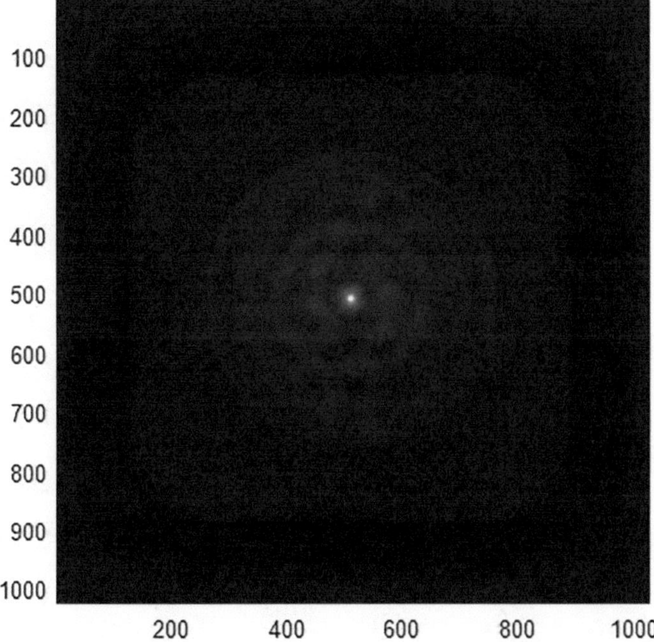

Fig. 1.19 Autocorrelation image corresponding to the direct autocorrelation of the sparse aperture shown in Fig. 1.14

aperture with an annulus of width $= 16$ pixels shown in Fig. 1.20b. Other sparse apertures with seven sub-apertures, each of radius 16 pixels and all filled with a circle of radius 256 pixels, are shown in Fig. 1.21. All speckle images have dimensions equal to 256×256 pixels, as shown in Figures 1.20 and 1.21.

The contrast is computed from equation (1.15) for the above speckle images shown in Figs. 1.20 and 1.21. The contrast values are obtained using the MATLAB code and are represented in equation (1.15). Referring to the contrast values obtained in Figs. 1.20 and 1.21, we found that the contrast changed from $K1 = 0.5112$ to $K3 = 0.5146$ when the seven sub-apertures filled the circle, giving better resolution. Compromised resolution and contrast are obtained using the annular aperture surrounding the sparse aperture. We have an improved contrast of values, $K2 = 0.5259$ and $K4 = 0.5256$, compared to the values of K in the absence of the annulus. In addition, the resolution is improved in the case of the annulus surrounding the sub-apertures.

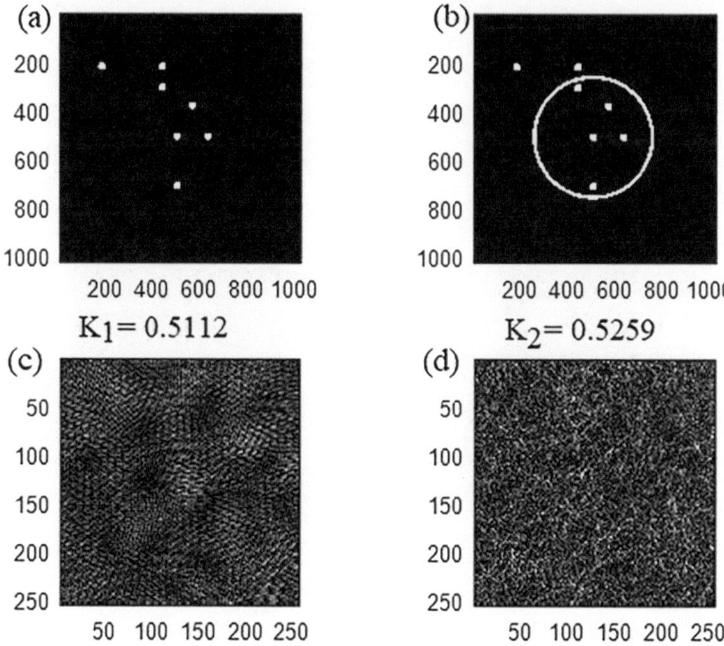

Fig. 1.20 The speckle image shown in **c** corresponds to the sparse aperture shown in **a**, and the speckle image shown in **d** corresponds to the sparse aperture with the annulus of width = 16 pixels shown in **b**. The sparse aperture is composed of seven sub-apertures, each of which has a radius of 16 pixels, and they are outside a circle of radius 256 pixels. The corresponding speckle contrasts are K1 and K2

Finally, the point array autocorrelation (PAA) is affected by the distribution of the sub-apertures and the annulus surrounding the sparse aperture configuration.

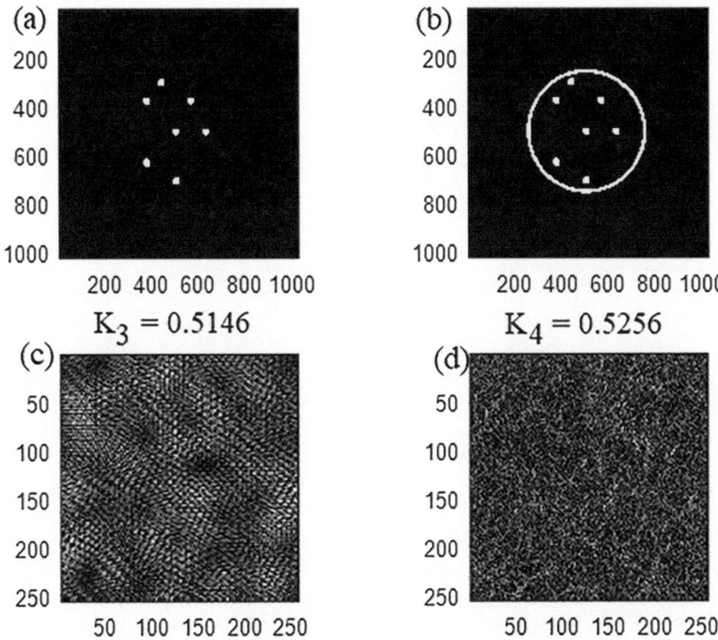

Fig. 1.21 The speckle image shown in **c** corresponds to the sparse aperture shown in **a** and the speckle image shown in Figure **d** corresponds to the sparse aperture with the annulus of width = 16 pixels shown in **b**. The sparse aperture has seven sub-apertures, each of which has a radius of 16 pixels and a filling circle with a radius of 256 pixels. The corresponding speckle contrasts are = K3 and K4

1.4 Conclusion

First, we computed the lateral PSF for the linear and quadratic distribution of the sub-apertures and compared them with that corresponding to the case of uniform circular sub-apertures. Five sub-apertures are considered using the FFT technique in the computation of the PSF. The intensity of the secondary peaks decreased with the increase of the sub-aperture diameter. Second, we investigated the sparse aperture imaging system modulated by an annulus surrounding the sub-apertures. We optimized the lateral PSF corresponding to the sparse aperture of 7 sub-apertures as shown in Fig. 1.16. We deduced that the bandwidth (BW) obtained from the lateral PSF is decreased for a full-filled aperture; hence the resolution is improved.

We deduced that the contrast of the speckle images is improved for the sub-apertures surrounded by an annulus. In addition, the microscope resolution, either conventional or confocal, is enhanced in the case of the annulus surrounding the sparse aperture.

Table 1.1 The first secondary peak on the lateral PSF around the central peak normalized to unity. The Golay-5 sub-apertures in the absence of the annulus are considered in Fig. 1.7

Radius of the sub-aperture (pixels)	The 1st secondary peak of the lateral PSF
16	0.942
32	0.8221
40	0.7391
48	0.6457
56	0.5436
64	0.4395

References

1. P.S. Salvaggio, J.R. Schott, D.M. McKeown, Genetic apertures: an improved sparse aperture design framework. Appl. Opt. **55**(12), 3182–3191 (2016). https://doi.org/10.1364/AO.55.003182
2. C. Zhou, Z. Wang, Mid-frequency MTF compensation of optical sparse aperture system. Opt. Express **26**(6), 6973–6992 (2018). https://doi.org/10.1364/OE.26.006973
3. N.J. Miller, M.P. Dierking, B.D. Duncan, Optical sparse aperture imaging. Appl. Opt. **46**(23), 5933–5943 (2007). https://doi.org/10.1364/AO.46.005933
4. F. Cassaing, L.M. Mugnier, Optimal sparse apertures for phased-array imaging. Opt. Lett. **43**(19), 4655–4658 (2018). https://doi.org/10.1364/OL.43.004655
5. A.J. Stokes, B.D. Duncan, M.P. Dierking, Improving mid-frequency contrast in sparse aperture optical imaging systems based upon the Golay-9 array. Opt. Express **18**(5), 4417–4427 (2010)
6. L.M. Mugnier, G. Rousset, F. Cassaing, Aperture configuration optimality criterion for phased arrays of optical telescopes. J. Opt. Soc. Am. A **13**(12), 2367–2374 (1996)
7. I. Tcherniavski, M. Kahrizi, Optimization of the optical sparse array configuration. Opt. Eng. **44**(10), 103201 (2005)
8. J.R. Fienup, J.J. Miller, Comparison of reconstruction algorithms for images from sparse-aperture systems. Proc. SPIE **4792**, 1–8 (2002)
9. H. Chen, Z. Cen, C. Wang, L. Shun, L. XiaoTong, Image restoration via improved wiener filter applied to optical sparse aperture systems. Optik **147**, 350–359 (2017)
10. H.U.I. Mei, W.U. Yong et al., Image restoration for synthetic aperture systems with a non-blind deconvolution algorithm via a deep convolutional neural network. Opt. Express **28**(7), 9929–9943 (2020)
11. A.M. Hamed, Resolution and contrast in confocal optical scanning microscope. Opt. and laser technology **16**, 93–96 (1984)
12. A.M. Hamed, Formation of speckle images for diffusers illuminated by modulated apertures (circular obstruction). J. Mod. Opt. **56**, 1633–1642 (2009). https://doi.org/10.1080/09500340903277792
13. A.M. Hamed, Discrimination between speckle images using diffusers modulated by some deformed apertures: Simulations. Opt. Eng. **50**, 1–7 (2011). https://doi.org/10.1117/1.3530085
14. A.M. Hamed, T. Al-Saeed, Image analysis of modified Hamming aperture: application on confocal microscopy and holography. J. Modern Opt. **62**, 801–810 (2015)
15. A.M. Hamed, Speckle imaging of annular Hermite Gaussian laser beam. Pram. J. Phys. **95**, 202 (2021). https://doi.org/10.1007/s12043-021-02231-9
16. Xie Zong Liang, Ma Haotong, et al., Restoration of sparse aperture images using spatial modulation diversity technology based on a binocular telescope testbed. IEEE Photonics J. **9**(3), 7802611 (2017)

Chapter 2
Investigation of Hexagonal Sparse Apertures Surrounded by a Rectangular Annulus

We designed an aperture of five hexagonal sparse apertures surrounded by an annular aperture. We computed the impulse response, also known as the point spread function (PSF), corresponding to the described aperture. We have obtained the autocorrelation corresponding to the hexagonal sparse apertures. The spatial resolution of the conventional and confocal microscopes, using the above apertures, is obtained. We compared the results with those corresponding to the uniform rectangular and annular rectangular apertures. We used hexagonal sparse apertures followed by a diffuser to create speckle images, and then reconstructed the aperture from the speckle pattern.

2.1 Introduction

Rectangular sub-apertures, in optics, are rectangular openings that selectively allow light to pass through while blocking other areas. They are used to define the boundaries of an optical path, mask out specific areas, or shape the beam profile. They can be used in various applications, including imaging systems, monochromators, and detectors [1].

Optimizing the structural parameters of a sparse aperture with four rectangular sub-apertures focuses on maximizing image quality, particularly by improving the modulation transfer function (MTF) in all directions. This involves finding the optimal configuration of the sub-apertures, specifically their length-to-width ratio, to achieve a more uniform MTF distribution and consistent imaging performance. Image simulation of the system reveals that the resolution of the unoptimized structure is low in both horizontal and vertical directions. The image quality of the optimized structure is similarly high in all directions [1, 2].

Sparse apertures can greatly improve the size, weight, power, and cost compared to larger multi-meter diameter monolithic apertures for Intelligence, Surveillance, and Reconnaissance (ISR) applications. However, their system design is more complex

A. M. Hamed, *Image Processing Techniques for Deformed and Sparse Aperture Systems*,
SpringerBriefs in Applied Sciences and Technology,
https://doi.org/10.1007/978-3-032-04921-6_2

and requires careful optimization of the aperture configuration. The Air Force Research Laboratory (AFRL) has developed a simulation toolkit that enables rapid prototyping of sparse aperture designs to evaluate their optical performance, which assists in concept design development [3]. Resolution-enhanced sparse aperture imaging is discussed in [2].

The authors select arrays constructed on the compact nonredundant point arrays described by Golay [4]. They reported the results of synthesizing an image from multiple sub-aperture pupil fields by masking a large lens with a Golay array, noting the contrast reduction inherent to sparse arrays [5].

We calculated the point spread function (PSF) corresponding to the black/white (B/W) hexagonal aperture in [6, 7] and extracted the resolution information from the cut-off spatial frequency. The effect on the PSF is quite noticeable, as energy is displaced from the central peak to the sidelobes [8].

In this chapter, we suggested hexagonal sparse apertures surrounded by a rectangular annulus. We computed the corresponding point spread function (PSF) and extracted the resolution information. In addition, we applied these apertures in speckle imaging and reconstructed the aperture using the FFT. We discussed the results and concluded.

2.2   Theoretical Analysis

In Chap. 1, we computed the point spread function considering five sparse circular apertures surrounded by an annulus, and we obtained the following:

$$
h(u, v) = \left[\frac{J_1(w)}{w} - \epsilon^2 \frac{J_1(\epsilon w)}{\epsilon w} \right] + \left[\frac{J_1\left(\frac{w}{16}\right)}{\frac{w}{16}} \right]
$$

$$
\left\{ 1 + \exp\left(-\frac{j2\pi}{\lambda f} x_0 u \right) + \exp\left[-\frac{j2\pi}{\lambda f} \left(\frac{x_0}{2} u + y_0 v \right) \right] \right. \tag{2.1}
$$

$$
\left. + \exp\left[\frac{j2\pi}{\lambda f} \left(\frac{3x_0}{2} u - y_0 v \right) \right] + \exp\left[\frac{j2\pi}{\lambda f} \left(\frac{3x_0}{2} u + y_0 v \right) \right] \right\}
$$

In Eq. (2.1), the PSF corresponding to the annulus of circular shape is as follows:

$$
h_{annul}(w) = \left[\frac{J_1(w)}{w} - \varepsilon^2 \frac{J_1(\varepsilon w)}{\varepsilon w} \right] \tag{2.2}
$$

$\varepsilon = 1 - \frac{\delta \rho}{\rho_0}$ represents the ratio of the internal to the external radius for the uniform circle of radius ρ_0, the reduced coordinate $w = \frac{2\pi}{\lambda f} \rho_0\, r$, and r is the radial coordinate in the Fourier plane:

$$h_{sparse\ cir}(w) = 2\left[\frac{J_1\left(\frac{w}{16}\right)}{\frac{w}{16}}\right] \tag{2.3}$$

Following a similar analysis, where the rectangular annulus replaces the circular annulus, and hexagonal sparse apertures replace the circular sparse apertures. We compute the PSF corresponding to the hexagonal sparse apertures in the next section and add the PSF corresponding to the rectangular aperture.

2.2.1 PSF of Hexagonal Sparse Apertures Surrounded by a Rectangular Annulus

We summarize the PSF results for the recently hexagonal sparse aperture [6] as follows:

$$P(x, y) = \text{rect}(x, y) + \text{tri}\left(x, y - \frac{2a}{3}\right) + \text{tri}\left(x, y + \frac{2a}{3}\right) \tag{2.4}$$

where $\text{rect}(x, y) = 1$; $\left|\frac{x}{\left(\frac{b}{16}\right)}\right| \leq 1$ and $\left|\frac{y}{\left(\frac{b}{16}\right)}\right| \leq 1$ represents a rectangle of sides a/16 and b/16, where $a = b$.

$\text{tri}\left(x, y - \frac{2a}{3}\right) = 1$. It represents a triangle of base 2b/16 and height a/16.

We applied the Fourier transform using Eq. (2.4), and we obtained the PSF corresponding to the hexagonal sparse aperture as follows [6, 7]:

$$\begin{aligned}
h_{sparse\ hex}(u, v) = &\left[\frac{\sin(\pi bu/16)}{(\pi bu/16)}\right]\left[\frac{\sin(\pi av/16)}{(\pi av/16)}\right] \\
&\left\{1 + 2\left[\frac{\sin(\pi bu/16)}{(\pi bu/16)}\right]\left[\frac{\sin(\pi av/16)}{(\pi av/16)}\right]\cos\left[\left(\frac{4\pi a}{3\lambda f}\right)(v/16)\right]\right\}
\end{aligned} \tag{2.5}$$

The PSF corresponding to the rectangular annulus is obtained as follows:

$$h_{annul\ rect}(w) = \left[\frac{\sin(\pi bu)}{(\pi bu)}\right]\left[\frac{\sin(\pi av)}{(\pi av)}\right] - \alpha\left[\frac{\sin(\pi b\prime u)}{(\pi b\prime u)}\right]\left[\frac{\sin(\pi a\prime v)}{(\pi a\prime v)}\right] \tag{2.6}$$

We have assumed that the external rectangle has sides of a and b, where $a = b$, and the internal sides of the rectangle are $a\prime = a - \Delta a$ and $b' = b - \Delta b$ where a' = b', and $\alpha = \frac{a'b\prime}{ab}$.

In this case, Eq. (2.1) for the PSF corresponding to the five sparse hexagonal apertures surrounded by a rectangular annulus is as follows:

$$\begin{aligned}
h(u, v) = &\left\{ \left[\frac{\sin(\pi bu)}{(\pi bu)} \right]\left[\frac{\sin(\pi av)}{(\pi av)} \right] - \alpha \left[\frac{\sin(\pi b'u)}{(\pi b'u)} \right]\left[\frac{\sin(\pi a'v)}{(\pi\pi a'v)} \right] \right\} \\
&+ \left[\frac{\sin(\pi bu/16)}{(\pi bu/16)} \right]\left[\frac{\sin(\pi av/16)}{(\pi av/16)} \right] \\
&\left\{ 1 + 2\left[\frac{\sin(\pi bu/16)}{(\pi bu/16)} \right]\left[\frac{\sin(\pi av/16)}{(\pi av/16)} \right]\cos\left[\left(\frac{4\pi a}{3\lambda f} \right)(v/16) \right] \right\} \\
&\left\{ 1 + \exp\left(-\frac{j2\pi}{\lambda f}x_0 u \right) + \exp\left[-\frac{j2\pi}{\lambda f}\left(\frac{x_0}{2}u + y_0 v \right) \right] \right. \\
&\left. + \exp\left[\frac{j2\pi}{\lambda f}\left(\frac{3x_0}{2}u - y_0 v \right) \right] + \exp\left[\frac{j2\pi}{\lambda f}\left(\frac{3x_0}{2}u + y_0 v \right) \right] \right\}
\end{aligned}$$

$$(2.7)$$

2.2.2 Coding and Decoding of Information Using Speckle Imaging

We employed a spatially filtered laser beam in the coding process. We used the hexagonal sparse aperture surrounded by a rectangular annulus as an object of complex amplitude A (x, y). The light transmitted is incident upon a diffuser d (x, y) placed in the object plane (x, y).

Hence, we represent the transmitted complex amplitude as follows:

$$T(x, y) = A(x, y).d(x, y) \tag{2.8}$$

The complex amplitude corresponding to the diffracted light is obtained by performing the Fourier transform on Eq. (2.8), and we have the following in the Fourier plane (u, v):

$$\tilde{T}(u, v) = F.T.\{A(x, y).d(x, y)\} = \tilde{A}(u, v) \otimes \tilde{d}(u, v) \tag{2.9}$$

$\tilde{A}(u, v) = F.T.\{A(x, y)\}$ represents the Fourier spectrum of the object or the PSF corresponding to the hexagonal sparse aperture surrounded by a rectangular annulus.

$\tilde{d}(u, v) = F.T.\{d(x, y)\}$ represents the Fourier spectrum of the diffuser, which produces another random pattern convoluted by the PSF to produce the modulated speckle image.

Consequently, the speckle image intensity is the modulus square of the diffracted complex amplitude represented by Eq. 2.9. It is written as follows:

$$I(u, v) = |\tilde{A}(u, v) \otimes \tilde{d}(u, v)|^2 \tag{2.10}$$

2.3 Results and Discussion

We constructed five hexagonal sparse apertures surrounded by a rectangular annulus, as shown in Fig. 1a. The total width of the external rectangle is 416 pixels, the annular rectangular width is 22 pixels, and the side length of the hexagonal sparse apertures is 24 pixels.

We computed the PSF or the amplitude impulse response corresponding to the aperture in Fig. 2.1a using Fast Fourier transform (FFT). The diffracted image or the intensity impulse response corresponding to the aperture shown in Fig. 2.1a is shown in Fig. 2.1b. We observed a central hexagonal diffracted beam followed by fringes along the edges of the aperture. Diffraction along the Cartesian coordinates due to the rectangular annular aperture is shown. We draw in Fig. 2.1c a normalized PSF plot versus the Cartesian coordinate along the x-direction. The pattern is not symmetric since the five sparse apertures are arranged randomly inside the rectangular aperture. The central peak is located at 70 pixels, surrounded by eight smaller peaks, and then decays randomly. The total range is 128 pixels.

A rectangular annulus, where the total width of the external rectangle = 416 pixels, and the annular rectangular width = 24 pixels, is plotted in Fig. 2.2a. A compared diffracted image or the intensity impulse response corresponding to the aperture shown in Fig. 2.2a is shown in Fig. 2.2b. We observed a symmetric diffraction pattern along the two Cartesian coordinates corresponding to the rectangular annular aperture. We draw in Fig. 2.2c a normalized PSF plot versus the Cartesian coordinate along the x-direction. The pattern is symmetric, and eight decreased peaks are observed around the central peak at 66 pixels, followed by decaying peaks. The total range is 128 pixels.

By comparing the intensity impulse responses in the case of the sparse hexagonal aperture surrounded by a rectangular annulus and the rectangular annular aperture, we observed these two remarks: The central peak in the case of the sparse aperture is shifted by 4 pixels when compared to the central peak corresponding to the rectangular annular aperture. In the second remark, we observed in the case of sparse hexagonal apertures, two peaks near the center, located at 66 pixels, have smaller values than other secondary peaks when compared to the regular decay for the rectangular annular aperture.

A uniform rectangular aperture, where the width of the rectangle = 416 pixels, is drawn in Fig. 2.3a for comparison with the hexagonal sparse aperture results. The diffracted image or the impulse response corresponding to the aperture is shown in Fig. 2.3a. We observed a symmetric diffraction pattern along the two Cartesian coordinates corresponding to the rectangular aperture. We plotted in Fig. 2.3c a normalized PSF plot versus the Cartesian coordinate along the x-direction. The pattern is symmetric, and the secondary decreased peaks are observed around the central peak at 66 pixels, as we expected. The total range is 128 pixels.

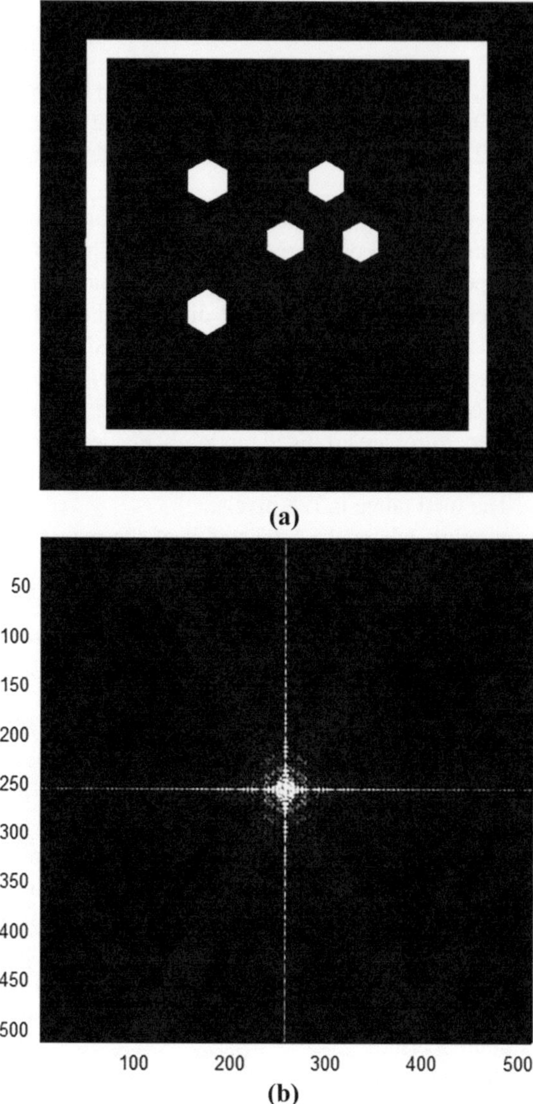

(a)

(b)

Fig. 2.1 **a** Five hexagonal sparse apertures surrounded by a rectangular annulus. The total width of the external rectangle = 416 pixels. The annular rectangular width = 22 pixels. The side length of the hexagonal sparse apertures = 24 pixels **b** The diffracted image or the impulse response corresponding to the aperture shown in **a**. We observed a central hexagonal diffracted beam followed by fringes along the edges of the aperture. Diffraction along the Cartesian coordinates due to the rectangular annular aperture is shown **c** Normalized PSF plot versus the Cartesian coordinate along the x-direction. The pattern is not symmetric; the central peak is shifted at 70 pixels, surrounded by eight smaller peaks, and then decays randomly. The total range is 128 pixels. The effect on the PSF is quite noticeable, as energy is displaced from the central peak to the sidelobes

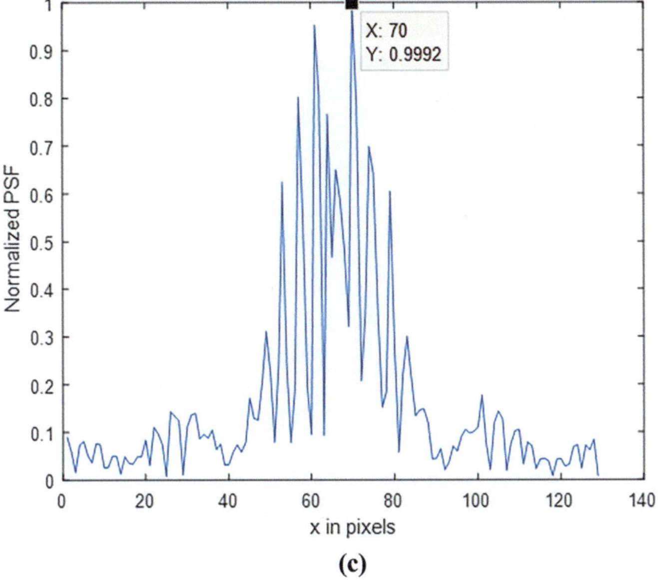

Fig. 2.1 (continued)

The total bandwidth corresponding to the hexagonal sparse apertures surrounded by a rectangular annulus, when compared to those corresponding to the annular rectangular and transparent rectangular apertures, has these values as shown in Table 2.1.

The spatial resolution of the conventional and confocal microscopes, using the above apertures, is extracted from the total bandwidth results. We observed that a smaller bandwidth in the spatial frequency domain has better resolution; hence, the hexagonal sparse aperture surrounded by a rectangular annulus has the optimum resolution. The corresponding total bandwidth $= 3$ pixels.

The autocorrelation image corresponding to the aperture shown in Fig. 2.1a is plotted in Fig. 2.4a. We observed a sharp central peak and eight other secondary peaks. The autocorrelation plots along both coordinates are shown clearly, as in Fig. 2.4b and c.

In Fig. 2.4b, the autocorrelation plot along the x-direction at a constant $y = 256$ pixels shows two secondary peaks with moderate values compared to the sharp central peak. They are about 38% of the central peak. The two secondary peaks are triangular, resulting from the autocorrelation of the rectangular annulus. Their widths are $160-112 = 48$ pixels and $396-352 = 44$ pixels, respectively.

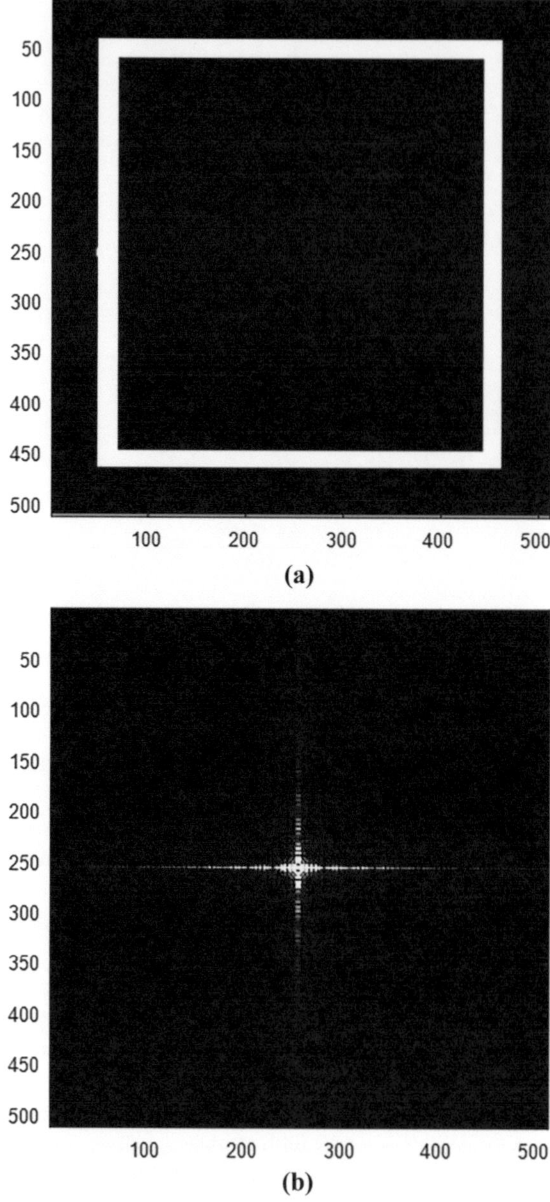

(a)

(b)

Fig. 2.2 **a** A rectangular annulus, where the total width of the external rectangle = 416 pixels, and the annular rectangular width = 24 pixels **b** The diffracted image or the impulse response corresponding to the aperture shown in **a**. We observed a symmetric diffraction pattern along the two Cartesian coordinates due to the rectangular annular aperture **c** We showed a Normalized PSF plot versus the Cartesian coordinate along the x-direction. The pattern is symmetric, and eight decreased peaks are observed around the central peak at 66 pixels, followed by decaying peaks. The total range is 128 pixels

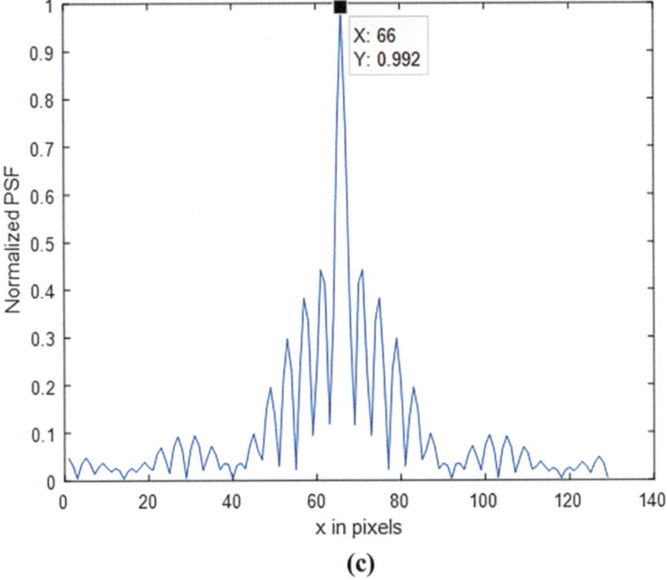

(c)

Fig. 2.2 (continued)

In Fig. 2.4c, the autocorrelation plot along the y-direction at constant x = 256 pixels. We showed four secondary peaks of moderate values compared to the central sharp peak. They are about 30% of the central peak. The two secondary peaks are triangular, resulting from autocorrelation of the rectangular annulus. Their widths are 169–131 = 38 pixels and 383–343 = 40 pixels, respectively.

In Fig. 2.5a, the autocorrelation image corresponding to the transparent rectangular aperture in Fig. 2.3a is plotted for comparison. The autocorrelation plot along the x-direction at constant y of 256 pixels is plotted in Fig. 2.5b. We showed a central peak of triangular shape, as expected for the uniform rectangular aperture.

Five sparse hexagonal apertures surrounded by a rectangular annulus are correlated with a transparent rectangular aperture. The cross-correlation along the x-direction at constant y of 256 pixels is plotted in Fig. 2.6a. We showed a central peak surrounded by secondary peaks in the form of a window, followed by a triangular shape.

Five sparse hexagonal apertures surrounded by a rectangular annulus are correlated with a transparent rectangular aperture. In this case, the cross-correlation is along the y-direction at constant x of 256 pixels as shown in Fig. 2.6b. We showed a central peak surrounded by secondary peaks in the form of a window, followed by a triangular shape, and then oscillating around 0.85.

A sparse hexagonal aperture surrounded by a rectangular annulus is considered an object. It is represented in Fig. 2.7a. It has dimensions of 2048 × 2048 pixels. The object is used to produce the speckle image using a certain diffuser of the same dimensions as the object.

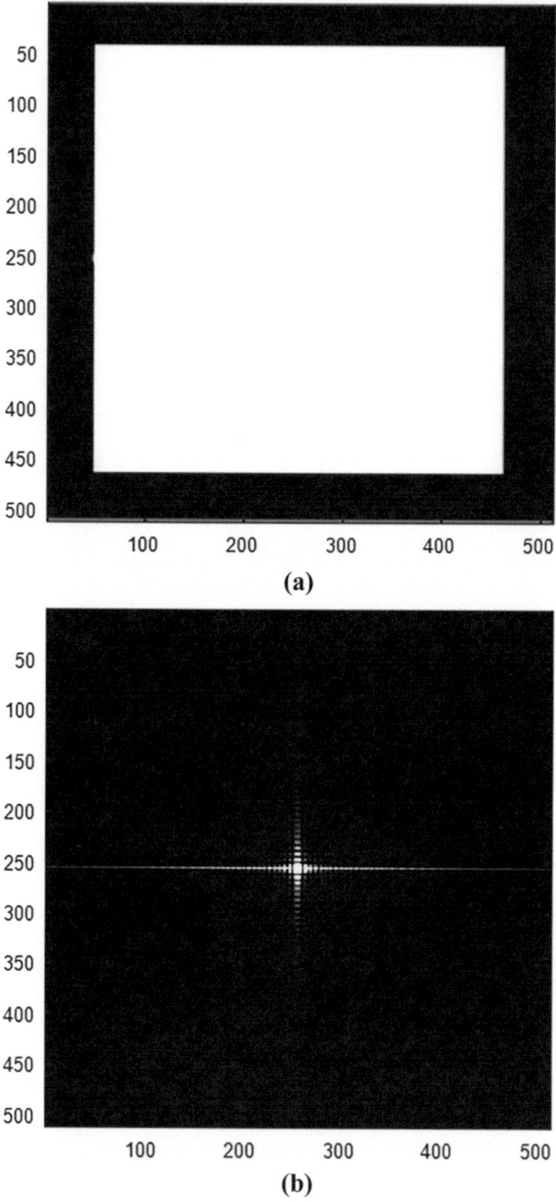

Fig. 2.3 **a** A uniform rectangular aperture, where the width of the rectangle = 416 pixels **b** The diffracted image or the impulse response corresponding to the aperture shown in **a**. We observed a symmetric diffraction pattern along the two Cartesian coordinates corresponding to the rectangular aperture **c** Normalized PSF plot versus the Cartesian coordinate along the x-direction. The pattern is symmetric, and the secondary decreased peaks are observed around the central peak at 66 pixels, as we expected. The total range is 128 pixels

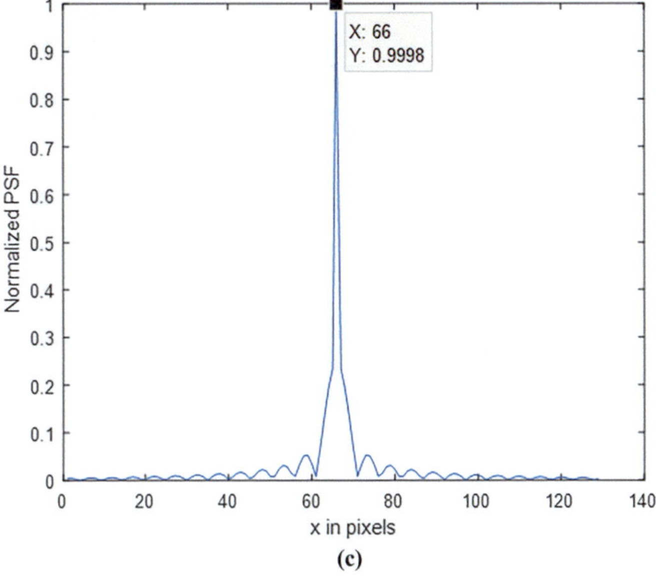

Fig. 2.3 (continued)

Table 2.1 The total bandwidth corresponding to three different apertures

Aperture	Total bandwidth in (pixels)
Hexagonal sparse aperture surrounded by a rectangular annulus	$72-69 = 3$
Rectangular annular aperture	$69-63 = 6$
Transparent rectangular aperture	$71-61 = 10$

The speckle pattern formed from the diffraction of the diffuser is multiplied by the object. The diffracted light from the object and the diffuser is realized using a converging lens. Digitally, we perform the FFT on the product of the transmitted amplitude from the object and the diffuser. The obtained speckle image has dimensions 512×512 pixels, as shown in Fig. 2.7b.

We reconstruct the object by performing the inverse Fourier transform upon the speckle image, and we obtain the image in Fig. 2.7c. We can improve the reconstructed image using Wiener filtering.

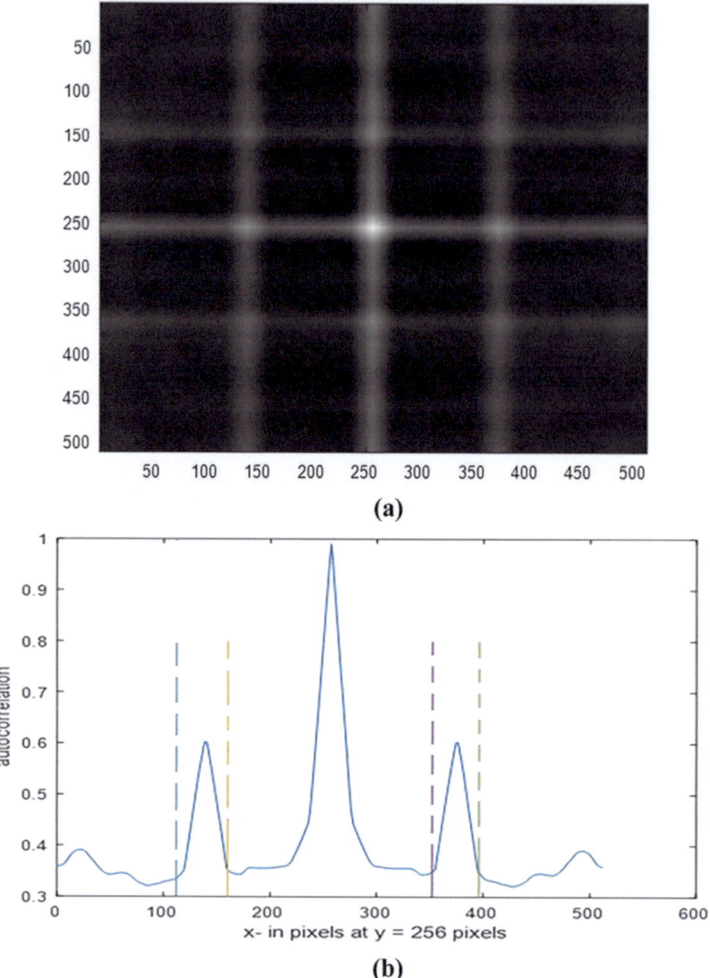

Fig. 2.4 **a** The autocorrelation image corresponding to the aperture shown in Fig. 2.1a. We observed a sharp central peak and eight other secondary peaks. This is shown clearly in the plots along both coordinates, as in **b** and **c**. **b** The autocorrelation plot along the x-direction at a constant y = 256 pixels shows two secondary peaks with moderate values compared to the sharp central peak. They are about 38% of the central peak. The two secondary peaks are triangular, resulting from the autocorrelation of the rectangular annulus. Their widths are 160–112 = 48 pixels and 396–352 = 44 pixels, respectively. **c** The autocorrelation plot along the y-direction at constant x = 256 pixels. We showed four secondary peaks of moderate values compared to the central sharp peak. They are about 30% of the central peak. The two secondary peaks are triangular, resulting from the rectangular annulus. Their widths are 169–131 = 38 pixels and 383–343 = 40 pixels, respectively

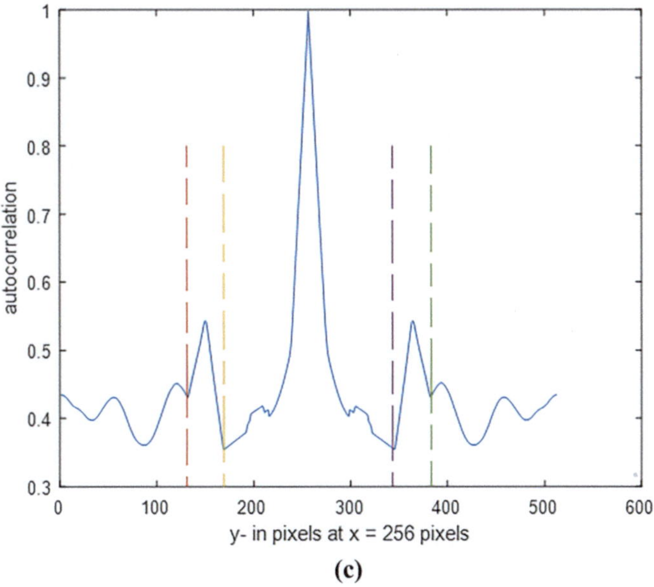

(c)

Fig. 2.4 (continued)

2.4 Conclusion

Referring to the PSF results, we concluded that the hexagonal sparse apertures surrounded by a rectangular annulus have a lower cut-off spatial frequency than the transparent rectangular and rectangular annular apertures.

The effect on the PSF corresponding to the hexagonal sparse apertures surrounded by a rectangular aperture is quite noticeable, as energy is displaced from the central peak to the sidelobes at 70 pixels, as shown in Fig. 2.1c.

From the autocorrelation results, we have obtained a sharp central peak and eight other secondary peaks corresponding to the hexagonal sparse apertures surrounded by a rectangular annulus. The two secondary peaks have moderate values of about 38% compared to the central peak.

Finally, we coded the aperture under investigation using a diffuser to obtain a speckle image. Then, we reconstructed the aperture by applying the inverse FFT.

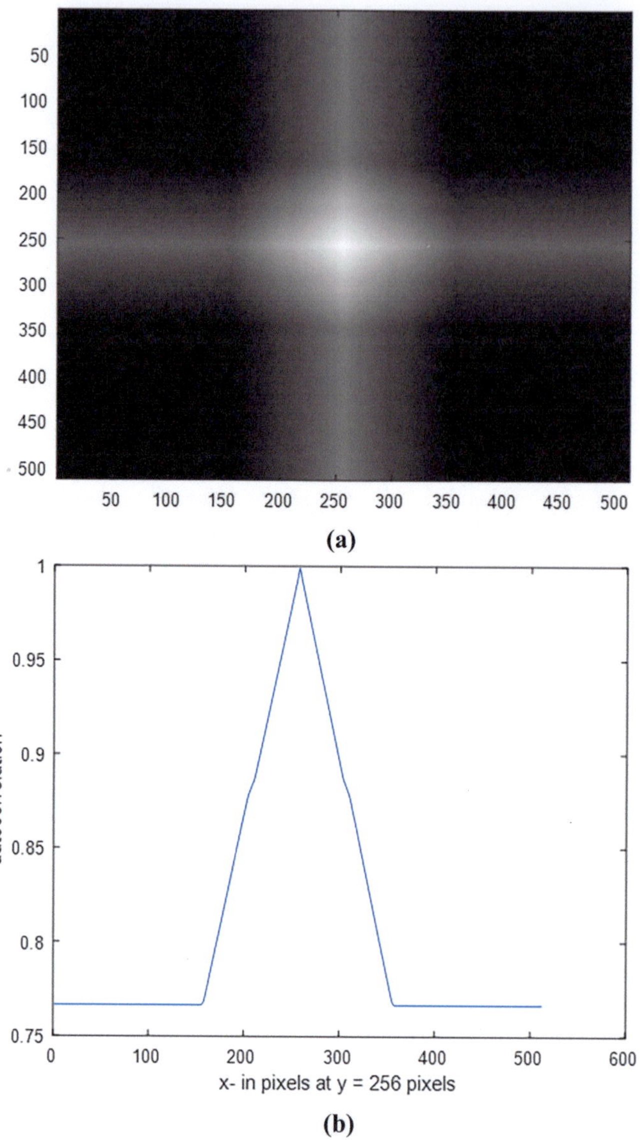

Fig. 2.5 **a** The autocorrelation image corresponding to the transparent rectangular aperture shown in **a**, **b** The autocorrelation plot along the x-direction at constant y = 256 pixels. We showed a central peak of triangular shape, as expected for the uniform rectangular aperture

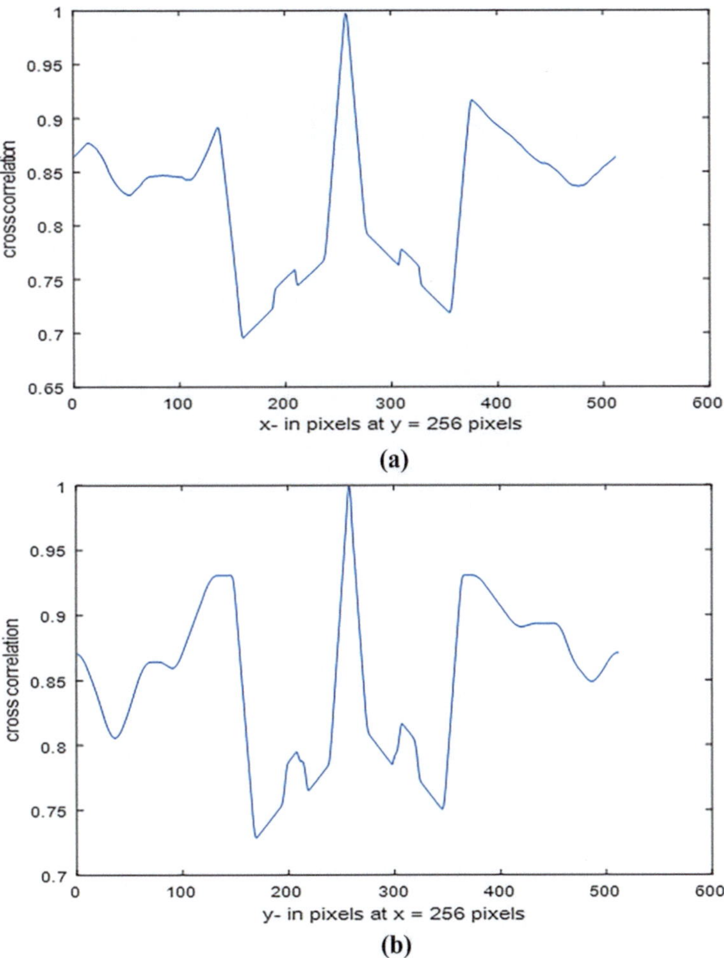

Fig. 2.6 **a** The cross-correlation plot along the x-direction at constant y = 256 pixels. We showed a central peak surrounded by secondary peaks in the form of a window, followed by a triangular shape. Five sparse hexagonal apertures surrounded by a rectangular annulus are correlated with a transparent rectangular aperture. **b** The cross-correlation plot along the y-direction at constant x = 256 pixels. We showed a central peak surrounded by secondary peaks in the form of a window, followed by a triangular shape, and then oscillating around 0.85. Five sparse hexagonal apertures surrounded by a rectangular annulus are correlated with a transparent rectangular aperture

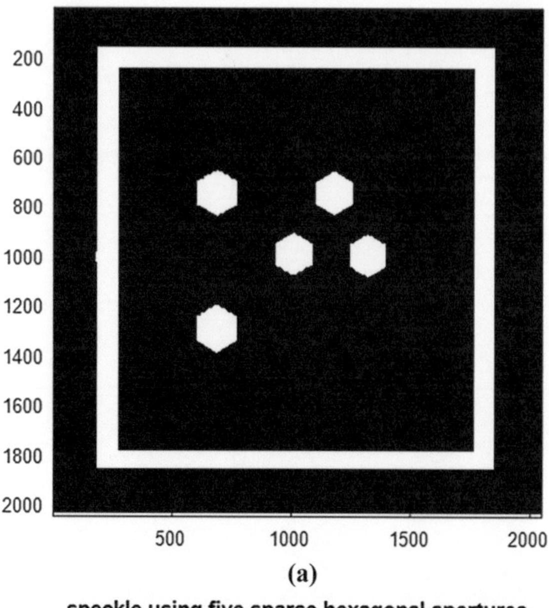

(a)

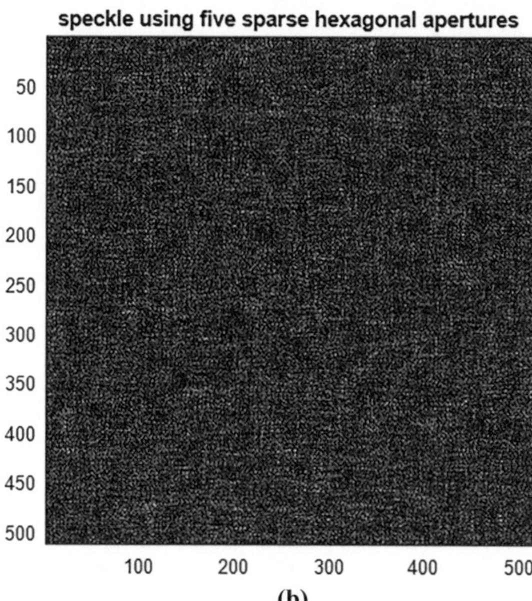

(b)

Fig. 2.7 **a** The object consists of sparse hexagonal apertures surrounded by a rectangular annulus. It has dimensions of 2048 × 2048 pixels. **b** The speckle pattern formed from the diffraction of the diffuser multiplied by the object. It has dimensions of 512 × 512 pixels. **c** The reconstructed image from the speckle image shown in **b**

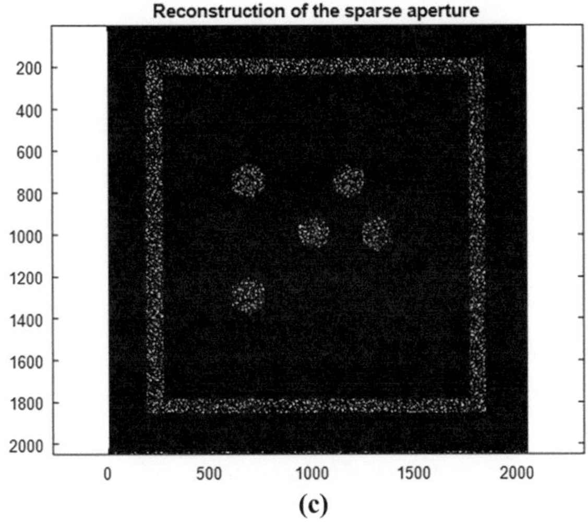

(c)

Fig. 2.7 (continued)

References

1. Z. Wang, Z. Zhao, and Li Liu, Optimization of structural parameters of sparse apertures with four rectangular sub-apertures. Conf. Proc., **11341** (2019). https://doi.org/10.1117/12.2543683
2. Joshua Snyder, Christopher Bailey, Steven Zuraski, Evaluating sparse aperture imaging performance: simulation and experiment. Proc. SPIE 12693, Unconventional Imaging, Sensing, and Adaptive Optics **1269303** (2023). https://doi.org/10.1117/12.2677525
3. N.J. Miller, B.D. Duncan, M.P. Dierking, Resolution enhanced sparse aperture imaging, in *Proceedings of IEEE Aerospace Conference. IEEE (2006) IEEEAC paper* 1406.
4. M.J.E. Golay, Point arrays having compact nonredundant autocorrelations. J. Opt. Soc. America **61**, 272–273 (1971)
5. N.J. Miller, M.P. Dierking et al., Optical sparse aperture imaging. Appl. Opt. **46**(23), 5933 (2007)
6. A.M. Hamed, Application of a hexagonal aperture on the confocal scanning laser microscope. Opt. Quant. Electron. **55**, 749 (2023)
7. A.M. Hamed, Studies on the confocal laser microscope, springer briefs in applied sciences and technology. (2025). https://link.springer.com/book/https://doi.org/10.1007/978-3-031-87275-4
8. S.-J. Chung, D. W. Miller, O.L. deWeck, Design and implementation of sparse aperture imaging systems, in Highly Innovative Space Telescope Concepts. H.A. MacEwen, ed., Proc. SPIE **4849**, 181–192 (2002)

Chapter 3
Design of Some Heterogeneous Apertures and Computation of the Resolution

3.1 Introduction

The confocal laser microscope consists of two objectives arranged in tandem, either in transmission or at reflection. The object in the common focus is mechanically scanned in synchronization with the electronic point detector located in the detection plane. This microscope was invented by M. Minsky [1]. After the invention of coherent sources, namely the laser, this microscope was manufactured by Davidovits et al. [2–4]. Early studies on this confocal microscope are presented by Sheppard et al. [5–8] and Wilson et al. [9, 10].

The modulated apertures are extensively investigated. Early, the annular aperture [1] was presented, followed by linear, quadratic, and higher order apertures presented by Hamed et al. [11–14]. The black and white (B/W) concentric annuli are investigated in [15, 16]. A recent publication on PSF corresponding to a linear-quadratic aperture is presented [17]. In addition, many biological and medical images are investigated using a confocal microscope with circular apertures [3, 18–20]. The application of modulated apertures in medical images using conventional and confocal microscopes is outlined in [21–28]. Also, the coherent transfer function (CTF) of confocal microscopes was computed.

The reconstructed images using the confocal microscope with the above-modulated apertures are presented in [26, 29].

In this chapter, three models of heterogeneous non-rotational symmetric apertures are selected. The first model has equal segments of circular and linear aperture, the second model has equal segments of circular and quadratic aperture, and the third model consists of equal linear and Hamming segments. PSF and the CTF are computed for these heterogeneous apertures. The CTF images corresponding to the heterogeneous apertures are obtained using the CLSM.

© The Author(s), under exclusive license to Springer Nature Switzerland AG 2025

A. M. Hamed, *Image Processing Techniques for Deformed and Sparse Aperture Systems*, SpringerBriefs in Applied Sciences and Technology, https://doi.org/10.1007/978-3-032-04921-6_3

3.2   Design of the Heterogeneous Rotational Apertures

3.2.1   First Model

The aperture has two equal segments of circular aperture, one half is transparent, while the other has a linear function.

We assumed that the upper half-circle is transparent while the lower half has a linear function. This heterogeneous aperture is mathematically represented as follows:

$$P_u(\rho) = 1 \text{ with} \left| \frac{\rho}{\rho_0} \right| \leq 1; 0 < \theta < \pi \tag{3.1}$$

$$P_l(\rho) = \rho \text{ with} \left| \frac{\rho}{\rho_0} \right| \leq 1; \pi < \theta < 2\pi \tag{3.2}$$

where the azimuthal coordinate is θ, ρ is the radial coordinate, and ρ_0 is the radius of the aperture. The whole heterogeneous aperture is the sum of the upper and lower segments of the transparent constant and the linear function, respectively.

Hence, the heterogeneous aperture is represented as follows:

$$P_{h1}(\rho) = P_u(\rho) + P_l(\rho) \tag{3.3}$$

$P_{h1}(\rho)$ stands for the heterogeneous aperture corresponding to the first model.

3.2.2   Second Model

The aperture has two equal segments of circular aperture, where one half is transparent while the other has a quadratic function.

We assumed the upper half-circle is transparent while the lower half has a quadratic function. This heterogeneous aperture is mathematically represented as follows:

$$P_u(\rho) = 1 \text{ with} \left| \frac{\rho}{\rho_0} \right| \leq 1; 0 < \theta < \pi \tag{3.4}$$

$$P_l(\rho) = \rho^2 \text{ with} \left| \frac{\rho}{\rho_0} \right| \leq 1; \pi < \theta < 2\pi \tag{3.5}$$

The whole heterogeneous aperture is the sum of the upper and lower segments of the transparent constant and the quadratic function, respectively.

Hence, the heterogeneous aperture is represented for the second model as follows:

$$P_{h2}(\rho) = P_u(\rho) + P_l(\rho) \tag{3.6}$$

$P_{h2}(\rho)$ stands for the heterogeneous aperture corresponding to the second model.

3.2.3 Third Model

The aperture has two equal segments in a circular aperture, where the upper half has a Hamming function while the lower half has a linear function.

Hence, this heterogeneous aperture is mathematically represented as follows:

$$P_u(\rho) = 0.54 + 0.46\cos(\beta\pi\rho); \text{ with} \left|\frac{\rho}{\rho_0}\right| \leq 1; 0 < \theta < \pi \tag{3.7}$$

β is a parameter, and its value is between zero and one:

$$P_l(\rho) = \rho \text{ with} \left|\frac{\rho}{\rho_0}\right| \leq 1; \pi < \theta < 2\pi \tag{3.8}$$

The whole heterogeneous aperture is the sum of the upper and lower segments of the Hamming and linear functions, respectively.

Hence, the heterogeneous aperture represents the third model as follows:

$$P_{h3}(\rho) = P_u(\rho) + P_l(\rho) \tag{3.9}$$

$P_{h3}(\rho)$ stands for the heterogeneous aperture corresponding to the third model.

3.3 Computation of the Point Spread Function (PSF) Corresponding to the Heterogeneous Apertures for the Three Models

The PSF for the upper transparent half circular segment is obtained using the Bessel-Fourier transform of Eq. (3.1) as follows:

$$h_{u1}(r) = \int_0^1 \int_0^\pi \exp\left[-j2\pi\left(\frac{\rho r}{\lambda f}\right)\cos(\theta)\right] \rho \, d\rho \, d\theta \quad \rho_0 = 1$$

$$= \pi \int_0^1 \rho \, J_0\left(\frac{k\rho r}{f}\right) d\rho \tag{3.10}$$

With $k = 2\pi/\lambda$ is the propagation constant.
The Eq. (3.10) is rewritten as follows:

$$h_{u1}(w) = \pi\left(\frac{f}{kr}\right) \int_0^W w J_0(w) dw \tag{3.11}$$

$W = k \, r/f$, and $w = \rho \, k \, r/f$ is a reduced coordinate. Finally, the PSF for the transparent circular segment was obtained as

$$h_{u1}(w) = \pi \frac{J_1(W)}{W} \tag{3.12}$$

The PSF for the lower linear segment is obtained using the Bessel-Fourier transform of Eq. (3.1) as follows:

$$h_{l1}(r) = \int_0^1 \int_\pi^{2\pi} \exp\left[-j2\pi\left(\frac{\rho r}{\lambda f}\right)\cos(\theta)\right] \rho \, d\rho \, d\theta \; ; \rho_0 = 1$$

$$= \pi \int_0^1 \rho^2 J_0\left(\frac{k\rho r}{f}\right) d\rho \tag{3.13}$$

Finally, the PSF for the linear segment was obtained as [11]

$$h_{l1}(W) = \pi\left[\frac{J_1(W)}{W} + \frac{J_0(W)}{W^2} - \frac{2\sum_i J_i(W)}{W^3}\right] \tag{3.14}$$

Finally, the PSF for the heterogeneous aperture shown in the first model is the sum of Eqs. (3.12) and (3.14), and we obtained this result:

$$h_{model1}(W) = 2\pi\left[\frac{J_1(W)}{W} + \frac{J_0(W)}{2W^2} - \frac{\sum_i J_i(W)}{W^3}\right] \tag{3.15}$$

An expression for the PSF is obtained for the second model, where the lower linear segment is replaced by a quadratic segment [17], giving this result:

$$h_{model2}(W) = 2\pi \left[\frac{J_1(W)}{W} - 2\frac{J_2(W)}{W^2} \right] \tag{3.16}$$

The PSF for the third model is the combination of the PSF for the upper Hamming segment and the lower linear segment, giving this result:

$$h_{model3}(W) = \pi \left[\frac{J_1(W)}{W} + \frac{J_0(W)}{W^2} - \frac{2\sum_i J_i(W)}{W^3} \right] + 0.54\,\delta(r) + \frac{1}{2}\left[\delta\left(r - \frac{\beta\lambda f}{2\rho_0}\right) + \delta\left(r + \frac{\beta\lambda f}{2\rho_0}\right) \right] \tag{3.17}$$

3.4 Effect of Rotation of the Heterogeneous Apertures

In this section, the considered heterogeneous apertures are rotated with an angle (θ) around their center in their plane (x, y). PSF is computed in all cases. If the aperture has a distribution P (x, y) represented by the described models shown above, then the rotated aperture is written as follows:

$$P_{rot.}(\rho) = P_0(\rho).\exp(ik\rho\,\sin\theta) \tag{3.18}$$

The PSF is computed for rotated apertures by applying the FFT techniques. It is known that the rotation corresponds to a translation in the Fourier plane.

3.5 Results and Discussion

The design of the first heterogeneous aperture is obtained from Eqs. (3.1, 3.2, and 3.3). A square matrix of dimensions 256×256 pixels is plotted, where the heterogeneous aperture has a radius $\rho_0 = 64$ pixels, the upper segment has a uniform transparent circle, while the lower has a linear function. This is plotted as shown in Fig. 3.1. The second and third models corresponding to C-Q and H–L are plotted as shown in Figs. 3.2, 3.3. The plots are obtained from Eqs. (3.4, 3.5, and 3.6) for the second model and Eqs. (3.7, 3.8, and 3.9) for the third model.

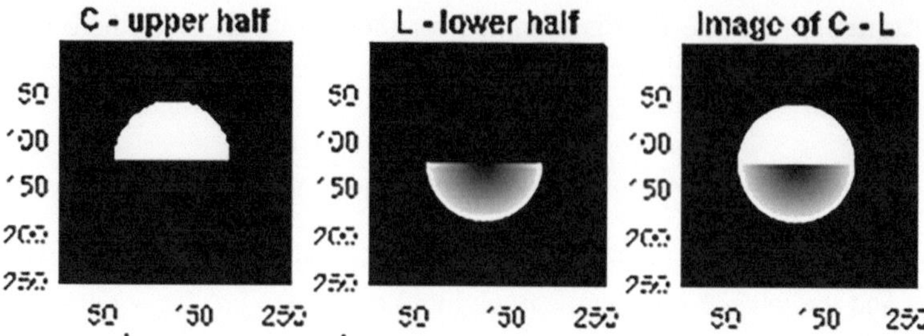

Fig. 3.1 On the left, a half-circular transparent aperture, in the middle, a linear half of the aperture, while on the right, the heterogeneous C-L aperture are shown. In the figure, L stands for the linear aperture while C stands for the uniform circular aperture. The aperture radius = 64 pixels

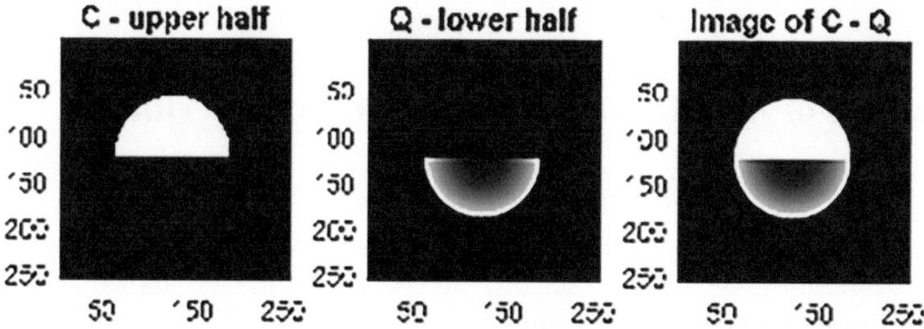

Fig. 3.2 On the left, a half-circular transparent aperture, in the middle, a quadratic function of half of the aperture, while on the right, the heterogeneous C-Q aperture are shown. In the figure, Q stands for the quadratic aperture while C stands for the uniform circular aperture. The aperture radius = 64 pixels

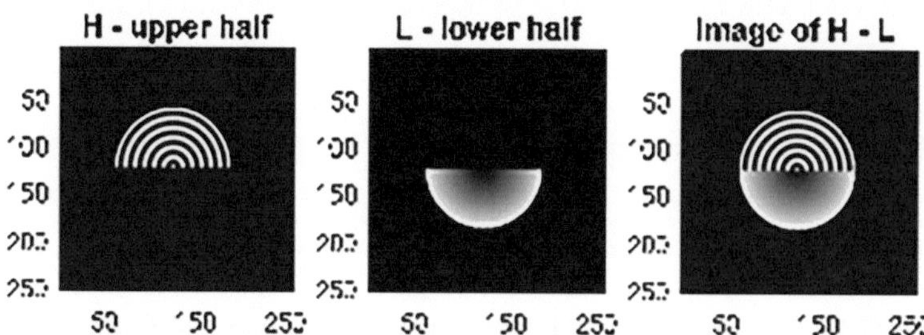

Fig. 3.3 On the left, the half-aperture of the Hamming function is shown, in the middle, the linear half of the aperture, while on the right, the heterogeneous H–L aperture are shown. In the figure, L stands for linear aperture while H stands for the Hamming aperture. The aperture radius = 64 pixels

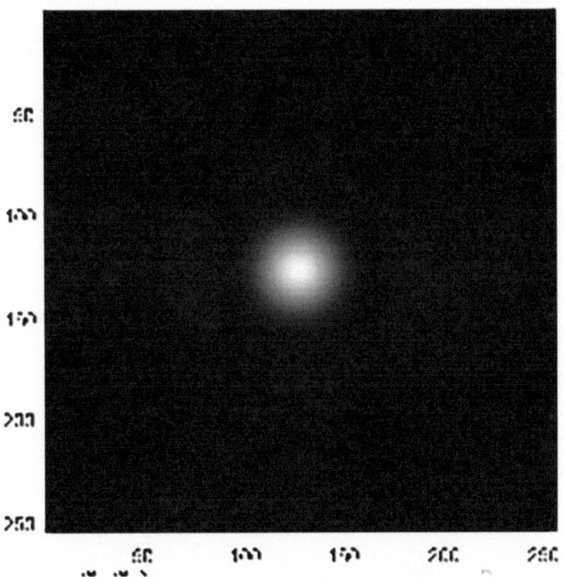

Fig. 3.4 The image of the PSF obtained using the model of the (C-L) aperture. The matrix of dimensions = 256 × 256 pixels, and the aperture radius = 4 pixels

The PSF image corresponding to the first model (C-L) of the heterogeneous aperture is plotted as shown in Fig. 3.4, using the FFT algorithm applied to the aperture. The aperture radius = 4 pixels. The line plot of the Normalized PSF for the first model taken at the center of the image at y = 128 pixels is shown in Fig. 3.5. In Fig. 3.5a, the aperture radius = 16 pixels and the corresponding cut-off spatial frequency at $r_c = 138\ pixels/\lambda f$. In Fig. 3.5b, the aperture radius = 8 pixels, and the corresponding cut-off spatial frequency $r_c = 148\ pixels/\lambda f$, while in Fig. 3.5c, the aperture radius = 4 pixels and the corresponding cut-off spatial frequency $r_c = 168\ pixels/\lambda f$. It is shown that the cut-off spatial frequency decreases from 168 to 138 pixels as the aperture radius is increased from 4 to 16 pixels. Hence, the resolution is improved using this model of C-L heterogeneous aperture using higher values of numerical aperture (NA).

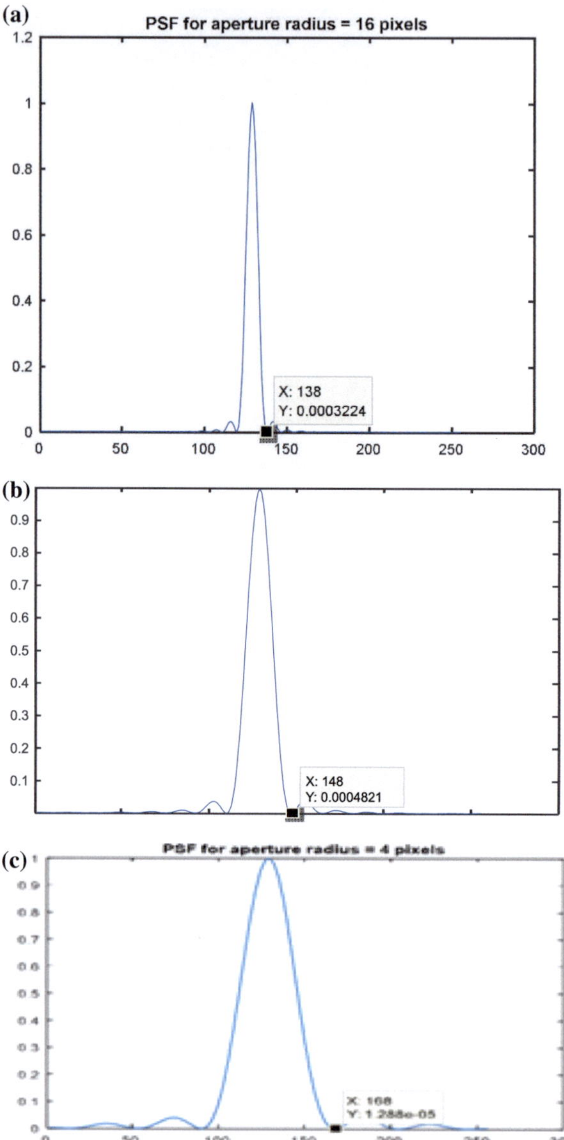

Fig. 3.5 **a** Normalized PSF for the aperture shown in the first model of the (C-L) aperture, where the aperture radius = 16 pixels. The cut-off spatial frequency equals 138 pixels/λf **b** Normalized PSF for the aperture shown in the first model of the (C-L) aperture, where the aperture radius = 8 pixels. The cut-off spatial frequency equals 148 pixels/λf **c** Normalized PSF for the aperture shown in the first model of the (C-L) aperture, where some aperture radius = 4 pixels. The cut-off spatial frequency = 168 pixels/λf

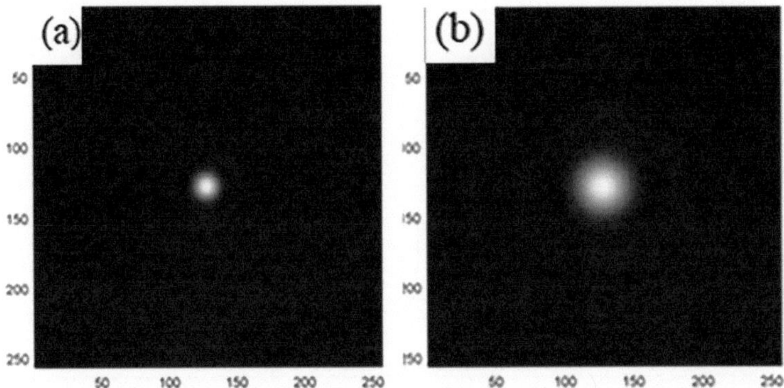

Fig. 3.6 The image of the PSF obtained using the second model of C-Q aperture. The matrix dimensions are 256 × 256 pixels, and the aperture radius is 8 pixels in (**a**), and 4 pixels in (**b**)

The PSF image for the second model (C-Q) is plotted as in Fig. 3.6a and b for aperture radius = 8, 4 pixels, respectively. The PSF line plot, at the center of the image, is presented in Fig. 3.7a, b, and c. Finally, the PSF image for the third model corresponding to the heterogeneous aperture (H–L) is plotted in Fig. 3.8 at radius = 16 pixels, and the PSF line plot at the center of the image is given in Fig. 3.9a, b, c where aperture radius = 16, 8, 4 pixels respectively.

The cut-off spatial frequency is unchanged for Models 2 and 3. Only the cut-off spatial frequency for low NA at radius 4 pixels is increased in the third model (H–L), giving r_c =173 pixels, which compares with r_c = 168 pixels corresponding to the first and second models. In addition, the fringing outside the central band is apparent in all models, which is useful in imaging extended objects.

The effect of the aperture rotation in its plane is outlined and represented by graphs. The PSF images for the C-Q aperture with radius = 8 pixels are shown in Fig. 3.10a. The left corresponds to the aligned aperture, while the right is the PSF for the inclined aperture at 45^0.

It is shown that the cut-off spatial frequency is changed from 145 pixels/λf to 177 pixels/λf (Fig. 10a). Hence, the resolution is decreased for the rotated heterogeneous aperture. If the aperture has a uniform distribution, its resolution remains constant while it rotates around itself as expected. In addition, the cut-off spatial frequency is changed from 146 pixels/λf to only 158pixels/λf for the third model of H–L aperture as shown in Fig. (3.11a, b). It is shown, referring to Fig. (3.11b, c, d), that as the angle of rotation increases, the cut-off spatial frequency increases, while the resolution decreases. Table 3.1 summarizes the PSF cut-off spatial frequency for the rotated H–L aperture.

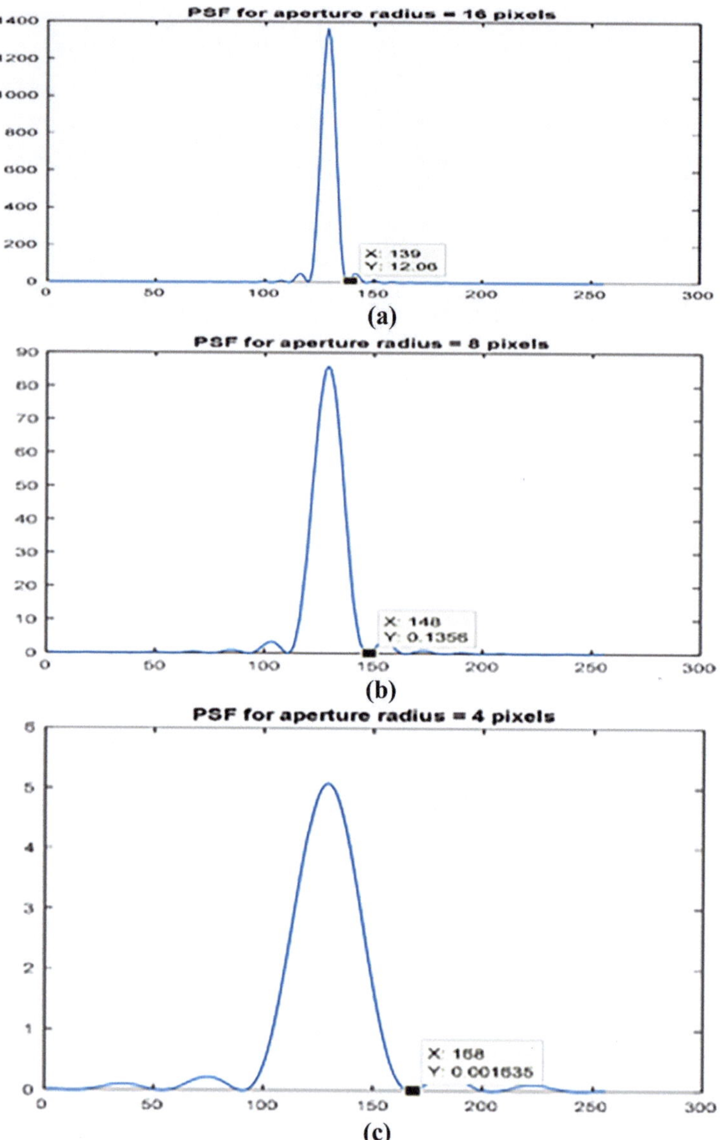

Fig. 3.7 **a** The line plot of PSF for the aperture shown in the second model of (C-Q) aperture, where the aperture radius = 16 pixels **b** The line plot of PSF for the aperture shown in the second model of (C-Q) aperture, where the aperture radius = 8 pixels **c** The line plot of PSF for the aperture shown in the second model of (C-Q) aperture, where the aperture radius = 4 pixels

Fig. 3.8 The image of the PSF obtained using the third model of (H–L) aperture. The aperture radius = 16 pixels

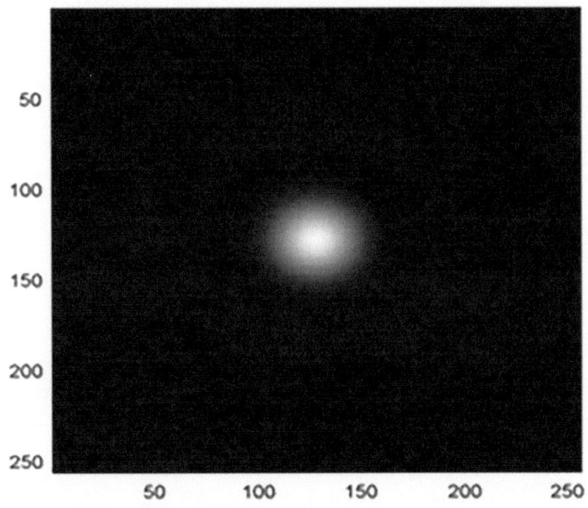

As shown in Fig. 3.12, the cut-off spatial frequency at $r_c = 161\ pixels/\lambda f$, $r_c = 165\ pixels/\lambda f$, and $r_c = 161\ pixels/\lambda f$ corresponding to rotation angles at $\theta = 60$, $75°$, and 90^0, respectively. For angle $\theta = 90°$, another cut-off at $181\ pixels/\lambda f$ is shown.

The CTF for H–L aperture is shown, where the upper plot is for two similar H–L apertures aligned, where the upper part has a Hamming distribution and the lower has a linear distribution. The next CTF has two rotated H–L apertures at $\theta = 45°$, while the third has one aligned and the other rotated by an angle $= 45°·$ The last CTF for the uniform circular aperture is given for comparison. All plots are shown in Fig. 3.13, computed using the FFT techniques.

It is shown, referring to the plots obtained, that the bandwidth of the Linear-Hamming (L–H) heterogeneous aperture of the CTF plots is much decreased for rotated H–L apertures as compared with the aligned apertures. This is attributed to the difference between correlated heterogeneous apertures during the rotation.

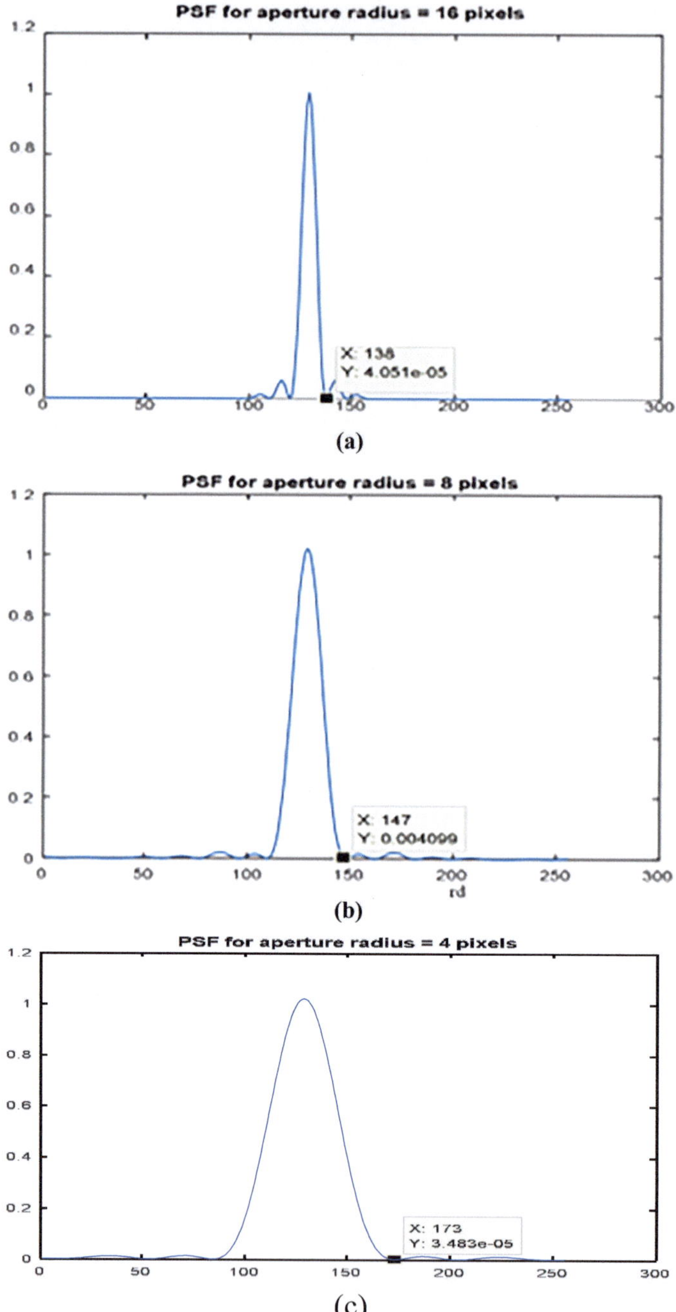

Fig. 3.9 PSF for the aperture shown in the 3 models of (H–L) aperture, where the aperture radius = 8 pixels shown in **b**. It is shown that the legs are weaker in intensity in the case of low NA compared with the strengthened legs shown for higher NA at a radius equal to 16 pixels shown in **a**. **c** PSF for the aperture shown in the third model of the (H–L) aperture. The aperture radius = 4 pixels

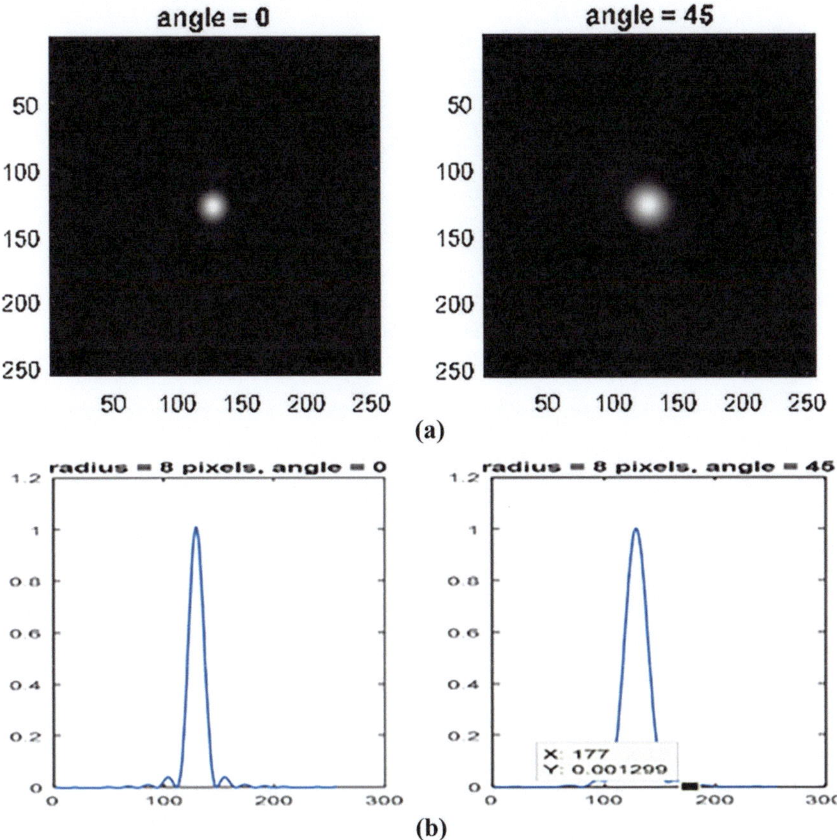

Fig. 3.10 **a** The PSF images for C-Q aperture with radius = 8 pixels, where the left corresponds to the aligned aperture, while the right is for the inclined aperture at 45^0 **b** Normalized PSF for the C-Q aperture. On the left, the PSF is shown for the aligned aperture, whereas on the right, the aperture is inclined at an angle $\theta = 45^0$

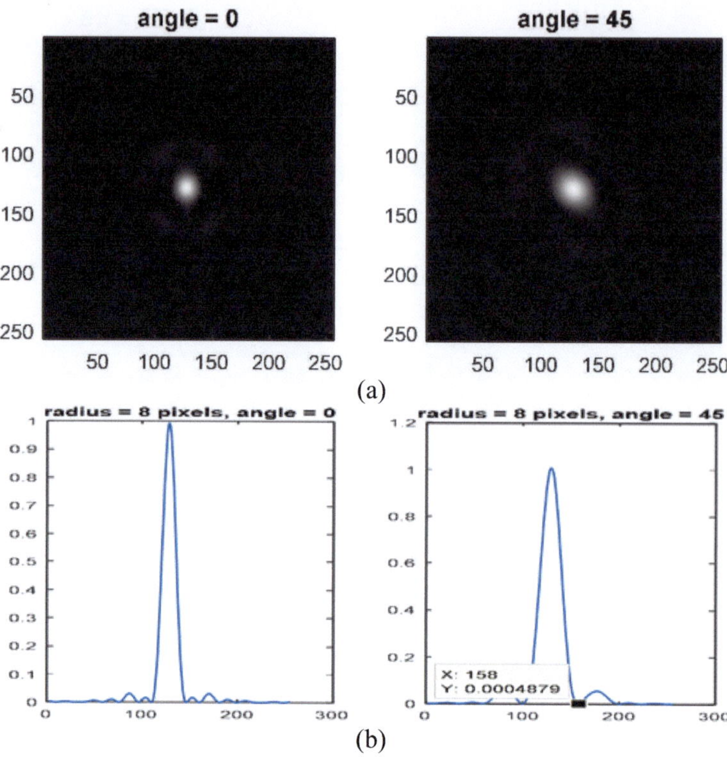

Fig. 3.11 **a** The PSF for H–L aperture, radius 8 pixels. In the left-aligned aperture, while in the right inclined aperture at 45^0 **b** Normalized PSF for the H–L aperture at radius 8 pixels. On the left, the PSF is shown for the aligned aperture, whereas on the right, the aperture is rotated at an angle $\theta = 45^0$. The cut-off spatial frequency for the rotated aperture $r_c = 158\,pixels/\lambda f$ compared with the aligned aperture, where $r_c = 147\,pixels/\lambda f$ **c** Normalized PSF for the H–L aperture at radius 8 pixels. On the left, the PSF is shown for the aligned aperture, whereas on the right, the aperture is rotated with angle $\theta = 30^0$. The cut-off spatial frequency for the rotated aperture $r_c = 156\,pixels/\lambda f$ compared with the aligned aperture where $r_c = 147\,pixels/\lambda f$ **d** Normalized PSF for the H–L aperture at radius 8 pixels. On the left, the PSF is shown for the aligned aperture, whereas on the right, the aperture is rotated with angle $\theta = 15^0$. The cut-off spatial frequency for the rotated aperture $r_c = 153\,pixels/\lambda f$ compared with the aligned aperture, where $r_c = 147\,pixels/\lambda f$

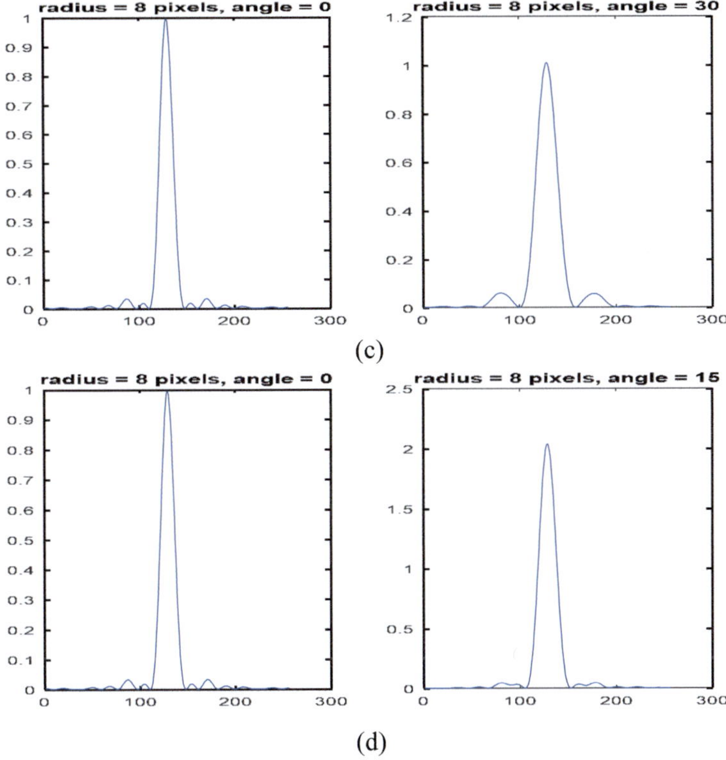

Fig. 3.11 (continued)

Table 3.1 Angle of rotation and the corresponding cut-off spatial frequency in the Linear-Hamming (L–H) heterogeneous aperture

The angle of rotation (θ)	Cut-off spatial frequency in pixels/λ f
0	147
15	153
30	156
45	158
60	161
75	164
90	161

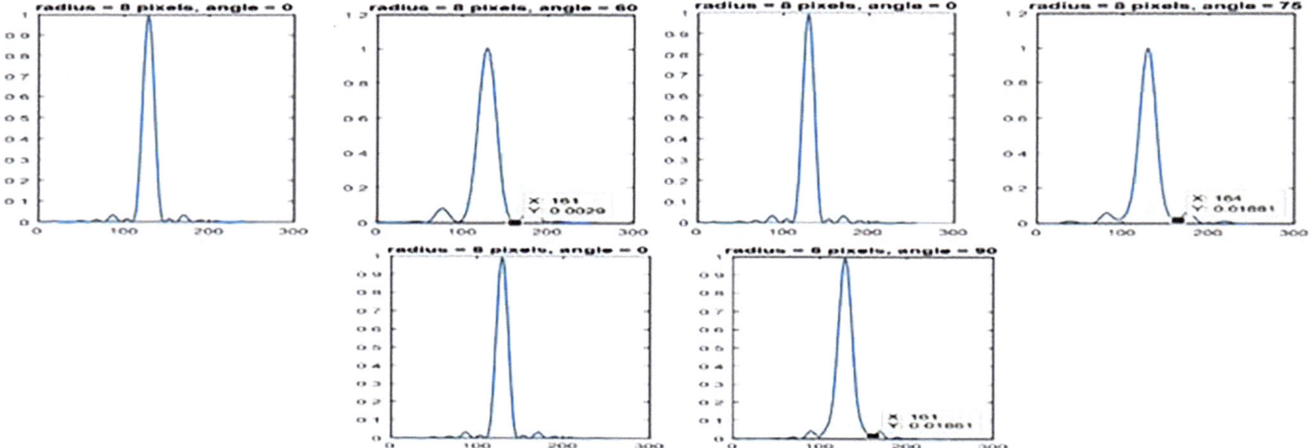

Fig. 3.12 Cut-off spatial frequency at $r_c = 161\frac{pixels}{\lambda f}$, $r_c = 165\frac{pixels}{\lambda f}$, *and* $r_c = 161\frac{pixels}{\lambda f}$ Corresponding to rotation angles of 60, 75, and 90, respectively. For angle 90, another cut-off at 181 $pixels/\lambda f$ is shown

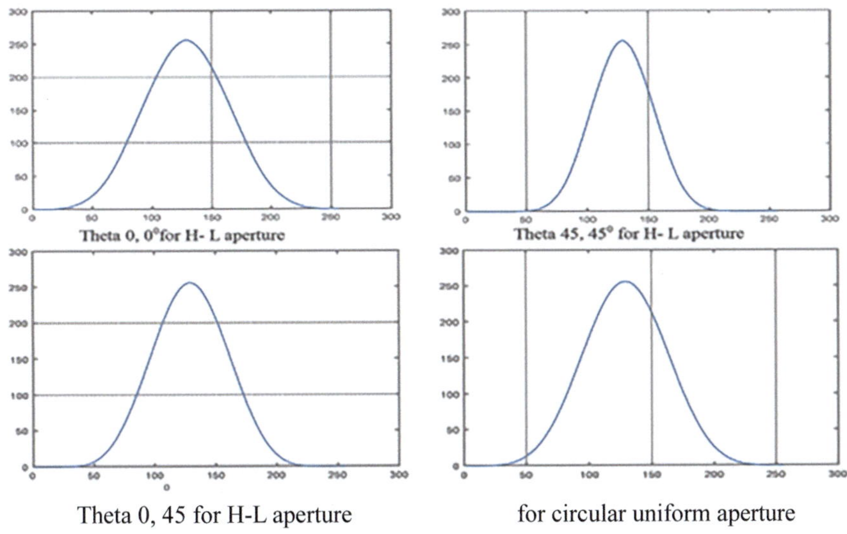

Theta 0, 45 for H-L aperture   for circular uniform aperture

Fig. 3.13 The CTF for the H–L aperture, where the upper plot is for two similar H–L apertures aligned. The upper part has a Hamming distribution, and the lower has a linear distribution. The next CTF has two rotated H–L apertures at $\theta = 45°$, while the third has one aligned and the other rotated by angle $\theta = 45°$. The last CTF for the uniform circular aperture is given for comparison

3.6 Conclusion

Recognition of aperture resolution is obtained from the cut-off spatial frequency computed from the PSF. The rotated heterogeneous apertures shown in the three models give different shapes of PSF when compared with the aligned apertures. An example is given for the third model of H–L aperture at constant radius = 8 pixels. Two plots are shown, one for the aligned aperture at $\theta = 0$ and the other for rotation at $\theta = 45°$. It is concluded that the cut-off spatial frequency is increased for the rotated aperture. Hence, to get the perfect alignment for this heterogeneous aperture (H–L), we rotate the aperture in its plane until we get the minimum cut-off spatial frequency. CTF is computed from the autocorrelation of the heterogeneous apertures, showing the difference between them and compared with the corresponding CTF for a circular aperture.

References

1. M. Minsky, 1957, U. S. Patent #3013467, Microscopy apparatus. Memoir on inventing the confocal scanning microscope, Scanning, **10**, 128–138 (1988)
2. P. Davidovits, M.D. Egger, Scanning laser microscope. Nature **223**, 831 (1969)
3. P. Davidovits, M.D. Egger, Scanning laser microscope for biological investigations. Appl. Opt. **10**(7), 1615–1619 (1971)

4. P. Davidovits, M.D. Egger, Scanning Optical Microscope. U.S. Patent #3,643,015 (1972)
5. C.J.R. Sheppard, A. Choudhury, Image formation in the scanning microscope. Optica **24**(10), 1051–1073 (1977)
6. I.J. Cox, C.J.R. Sheppard, Information capacity and resolution in an optical system. J. Opt. Soc. Am. A **3**(8), 1152–1158 (1986)
7. C.J.R. Sheppard, J.N. Gannaway, D. Walsh, T. Wilson, Scanning optical microscope for the inspection of electronic devices. In Proc. Microcircuit Engineering Conference, Cambridge University Press, 447–454, Cambridge (1978)
8. T. Wilson, C.J.R. Sheppard, *Theory and Practice of Scanning Optical Microscopy* (Academic Press, London, 1984)
9. T. Wilson, J.N. Gannaway, P. Johnson, A scanning optical microscope for inspection of semiconductor materials and devices. J. Microscopy **118**(3), 390–314 (1980)
10. T. Wilson, *Confocal Microscopy* (Academic Press, London, 1990)
11. A.M. Hamed, Numerical speckle images formed by diffusers using modulated conical and linear apertures. J. of Mod. Opt. **56**(10), 1174–1181 (2009)
12. A.M. Hamed, Formation of speckle images formed for diffusers illuminated by modulated apertures (circular obstruction). J. of Mod. Opt. **56**(15), 1633–1642 (2009)
13. A.M. Hamed, Discrimination between speckle images using diffusers modulated by some deformed apertures: Simulations. Opt. Eng. **50**(1), 1–7 (2011)
14. A.M. Hamed, Computer-generated quadratic and higher-order apertures and their application on numerical speckle images. Opt. and Photonics. J. **1**(2), 43–51 (2011)
15. A.M. Hamed, Scanning holography using a modulated linear pupil: simulations. Opt. and Photonics. J., **1** (2), 52–58, 2(011)
16. A.M. Hamed, Study of graded index and truncated apertures using speckle images. Precision. Inst. and Mech. **3**(1), 144–152 (2014)
17. A.M. Hamed, Improvement of point spread function (PSF) using linear-quadratic aperture. Optik **131**, 838–849 (2017)
18. G.J. Brakenhoff, E.A. Van Spronsen, H.T.M. Van der Voort, N. Nanning, Three-dimensional confocal fluorescence microscopy. Methods Cell Biol. **30**, 379–398 (1989)
19. W.B. Amos, J.G. White, M. Fordham, Use of confocal imaging in the study of biological structures. Appl. Opt. **26**(16), 3239–3243 (1987)
20. G.J. Brakenhoff, H.T.M. Van der Voort, E.A. Van Spronsen, N. Nanninga, Three-dimensional imaging by confocal scanning fluorescence microscopy. In: Recent Advances in Electron and Light Optical Imaging in Biology and Medicine (A. Somlyo, ed.), Ann. N. Y. Acad. Sci., New York, **483**, 405–414 (1986)
21. A.M. Hamed, Discrimination between normal and diseased stomach using speckle imaging. Int. J. Innovative Res. in Eng. and Management., **3**(2), 125–133 (2016)
22. A.M. Hamed, Investigation of SIDA Virus (HIV) images using interferometry and speckle techniques. Int. J. Inn. Res. in Comp. Sci. Tech., **4**, 38–45 (2016)
23. A.M. Hamed, Image processing of coronavirus using interferometry. Opt. and photonics J. **6**(5), 75–86 (2016)
24. A.M. Hamed, A Compromising of resolution and contrast using quadratic aperture in scanning holographic imaging. Int. J. of Photo. And Opt. Tech., **2**(2), 18–23 (2016)
25. A.M. Hamed, T.A. Al-Saeed, Reconstruction of the corneal layers affected by a periodic noise, Application on microscopic interferometry. Int. J. of Photo. And Opt. Tech., **2**(3), 6–12 (2016)
26. A.M. Hamed, S.Y. El-Zaiat, T. Al-Saeed, L. Hammad, Point spread function using longitudinal black and white stripes inside a circular aperture. Int. J. of Photonics and Opt. Tech., **3**(1) 1–9 (2017)
27. A.M. Hamed, A modified Michelson interferometer and an application on microscopic imaging". Int. J. of Photonics and Opt. Tech. **3**, 1–5 (2017)

28. A.M. Hamed, Processing of the retinal artery image using higher orders of two-beam interference. Int. J. of Photonics and Opt. Tech., **3**(2), 21−27 (2017)
29. Abdallah, Mohamed, Hamed, *The PSF of some modulated apertures* (Lambert Academic Publishing, Application on speckle and interferometry images), 2017)

Chapter 4
Study of Cascaded Black–Quadratic Aperture (CBQA)

A new Cascaded Black–Quadratic Distribution (CBQD) in circular aperture is suggested. Two different models of CBQD are studied. In the first model, ten strips are considered, where half are black and the other half have a quadratic distribution, starting with a black strip from the center of the circular aperture. In the second model, the annular aperture of the quadratic distribution is considered. We have computed the Point Spread Function (PSF) corresponding to the two arrangements and compared it with the corresponding PSF for different circular, annular, and black and white (B/W) transparent circular apertures. The cut-off spatial frequency, which is related to the resolution, is investigated in all the apertures described. In the confocal laser scanning microscope, we considered either of the described models in the formation of images.

4.1 Introduction

The principle of confocal microscopy was patented by Marvin Minsky in 1957, but it took several years before it was fully developed to incorporate a laser scanning process. The technique essentially scans an object point by point using a focused laser beam to enable three-dimensional reconstruction. In a conventional microscope, you can only see as far as the light can penetrate, whereas a confocal microscope images one depth level at a time. The Coherent Laser Scanning Microscope (CLSM) works by passing a laser beam through a light source aperture, which is then focused by an objective lens into a small area on the surface of your sample, and an image is built up pixel by pixel by collecting the emitted photons from the fluorophores in the sample. Consequently, in a confocal microscope, the object is illuminated with the focused image of a point source, and the reflected (or transmitted) light intensity is measured with a point detector focused onto the same point of the sample. In practice, the point source is achieved using laser illumination or an incoherent source with a

A. M. Hamed, *Image Processing Techniques for Deformed and Sparse Aperture Systems*,
SpringerBriefs in Applied Sciences and Technology,
https://doi.org/10.1007/978-3-032-04921-6_4

pinhole. The point detector is achieved using a pinhole in front of the detector. The intensity distribution is the modulus square of the convolution product of the object's complex amplitude and the Resultant Point Spread Function (RPSF) corresponding to the CLSM [1–10].

The CLSM is based on a conventional optical microscope, but instead of a lamp, a laser beam is focused on the sample [11, 12]. The intensity of the laser light is adjusted by neutral density filters and brought to a set of scanning mirrors that can move them precisely and quickly. One mirror tilts the beam in the X direction, the other tilts the beam in the Y direction. Together, they tilt the beam in a raster fashion or make mechanical scanning. The beam is then directed to the back focal plane of the objective lens and focused onto your sample. The effect of the scanning mirrors on this light is to produce a spot of non-scanned light, but standing still. This light then passes through a semi-transparent mirror, which takes it away from the laser and toward the detection system.

A lot of work on conventional and confocal microscope resolutions is presented using aperture modulation. For example, linear, quadratic, higher order apertures, graded index apertures, and other combinations of black and white (B/W) concentric transparent annuli were investigated in [13–17].

In this chapter, two models of the cascaded black and quadratic distributions (CBQD) inside the circular aperture were investigated, and the PSF was computed, getting resolution information from the cut-off spatial frequency. Reconstruction of microscopic images was attained using the CBQD apertures placed in front of the microscope objectives in the CLSM. The PSF results using the CBQD are compared with those corresponding to the PSF obtained in the cascaded black and linear distribution (CBLD) [17].

4.2   Theoretical Analysis

The first model of the CBQD aperture contains 10 equal zones starting from the center, which are black. Hence, only five strips of quadratic distribution are effective in the computation of the PSF. The aperture is shown as in Fig. 4.1 a and mathematically represented as follows:

$$P(u, v) = \sum_{j=1}^{N} \left\{ P_{2j}\left[2j \left(\frac{\rho}{\rho_0} \right)^2 \right] - P_{2j-1}\left[(2j-1)\left(\frac{\rho}{\rho_0} \right)^2 \right] \right\} \qquad (4.1)$$

where $\rho = \sqrt{u^2 + v^2}$ is the radial coordinate in the aperture plane of orthogonal coordinates (u, v), ρ_0 *is the total radius*, and N represents the total number of transparent segments. We assumed equal width for the black and quadratic strips. The relation between the coordinates (u, v) and the radial coordinate ρ is governed by this

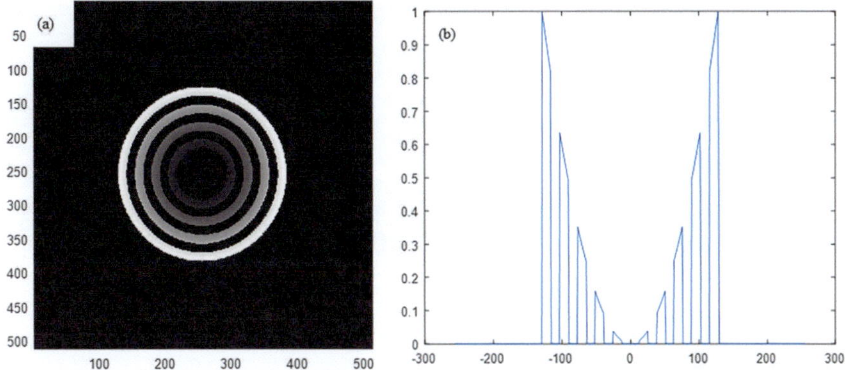

Fig. 4.1 In **a**, a Cascaded Black–Quadratic Distribution (CBQD) in a circular aperture is shown. Five zones have a quadratic distribution cascaded with another five dark zones, starting from the center dark zone. In **b**, a line plot corresponding to the described CBQD aperture is shown. The aperture radius = 128 pixels

transformation: $u = \rho\,cos(\theta)\ and\ v = \rho\,sin(\theta)$. The angle θ represents the azimuthal coordinate.

In this model, we take N = 5 for the five quadratic distributed segments.

The Point Spread Function (PSF) is computed by applying the Fourier transform to Eq. (4.1), and the PSF for the first model is obtained as follows:

$$h_{1st\ model}(Z) = \sum_{j=1}^{N}\left\{\left[\frac{J_1(\alpha_j Z)}{(\alpha_j Z)} - 2\,\frac{J_2(\alpha_j Z)}{(\alpha_j Z)^2}\right] - \left[\frac{J_1(\alpha_{j-1} Z)}{(\alpha_{j-1} Z)} - 2\,\frac{J_2(\alpha_{j-1} Z)}{(\alpha_{j-1} Z)^2}\right]\right\}$$

$$(4.2)$$

where $Z = \left(\frac{2\pi}{\lambda f}\right)\rho_0 r$. It is the reduced coordinate in the Fourier plane.

In the Eq. (4.2), $\alpha_j = \frac{2j}{N}$, and $\alpha_{j-1} = \frac{2j-1}{N}$, j = 1, 2..., N/2 and N = 10 represent the total number of zones while N/2 represents the total number of quadratic zones. Equation 4.2 is the PSF for quadratic and black strips of equal width, whatever the number of concentric annuli N.

4.3 Results and Discussion

The aperture composed of ten cascaded black and quadratic distributed functions described in the first model is drawn as in Fig. 4.1a, where equal strips are assumed. The image has dimensions of 512×512 pixels, while the aperture radius = 128 pixels. Five zones have quadratic distribution cascaded with another five dark zones starting from the center. The Line plot of the described CBQD aperture is shown in

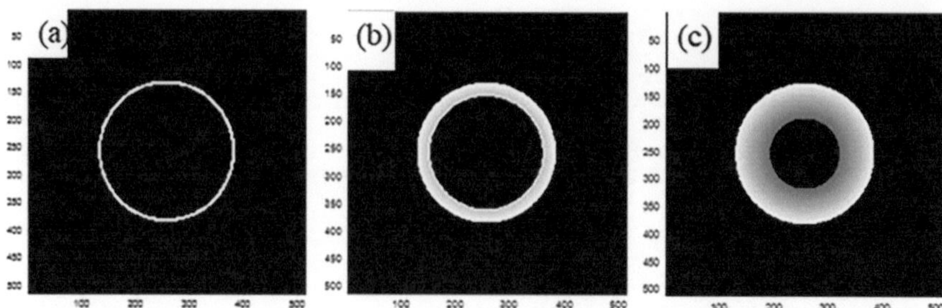

Fig. 4.2 In **a**, a linearly distributed annular aperture of width $= 0.05$ is shown, in **b**, a linear annulus of moderate width $= 0.2$ is shown, and in **c** a two-layer thick annulus of aperture ratio 1:1 is shown

Fig. 4.1b, where ten zones are considered, and five of them have quadratic distribution cascaded with another five black zones.

In the 2nd model, the annular aperture of linear distribution for different widths is shown in Fig. 4.2a–c. The annular width $\delta \rho = 0.05$ as in Fig. 4.2a, $\delta \rho = 0.2$ as in Fig. 4.2b, while the two-layer thick annulus of aperture ratio 1:1 is shown as in Fig. 4.2c).

In Figs. 4.3 and 4.4, we plotted the PSFs for the two models of CBQD apertures and compared them with the (CBLD) models at different radii represented in Table 4.1. The cut-off spatial frequency computed from the reduced coordinate is extracted from the curves, and the maximum amplitude at radius $= 16$ pixels is shown.

In Fig. 4.3a and c, in the first model of CBQD, we showed the optimum cut-off spatial frequency compared to the other models, except for the linear annulus model of width $= 0.05$. Referring to the results of the cut-off spatial frequency corresponding to all models shown in Table 4.1, we write this inequality:

$$W_c(linear\ annulus) = 0.6636 < W_c(CBQD) = 0.762 < W_c(CBLD)$$
$$= 0.8111 < W_c(CBWD) = 0.9586$$

It is shown, referring to the plots in Fig. 4.3c and d, that the cut-off spatial frequency corresponding to the first model of CBQD has the optimum resolution since $W_c = 0.762$ compared to the (CBLD) and cascaded B/W distribution (CBWD). An exception is the linear annulus, since the resolution is increased while the contrast is decreased.

It is shown that the legs of the diffraction pattern in the PSF for the suggested models are more stringent than the legs for the uniform circular aperture, which is useful for imaging extended objects.

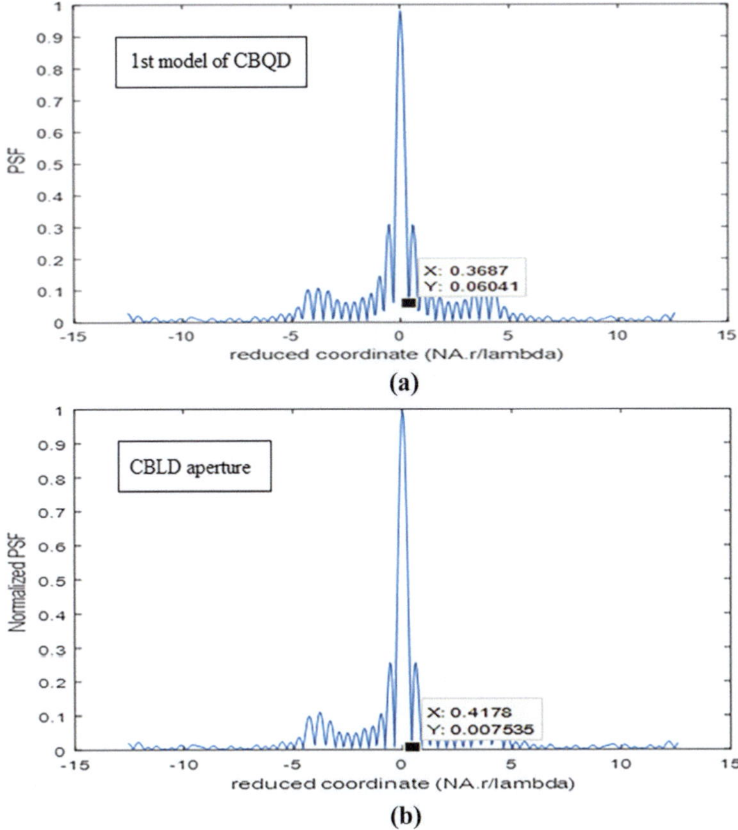

Fig. 4.3 **a** Normalized PSF for the CBQD aperture shown in Fig. 4.1, where the maximum radius = 32 pixels and the total number of zones N = 10. The cut-off spatial frequency in reduced coordinates is = 0.36870. **b** Normalized PSF for the CBLD aperture, where the maximum radius = 32 pixels and the total number of zones N = 10. The cut-off spatial frequency in reduced coordinates is = 0.4178. **c** The PSF versus the reduced coordinate = $NA.\frac{r}{\lambda}$. This corresponds to the model shown in Fig. 4.1, where the radius is 16 pixels. $PSF_{max.} = 224$ while the cut-off spatial frequency in reduced coordinates is = 0.762 **d** The PSF where the radius is = 16 pixels. $PSF_{max.} = 291$ while the cut-off spatial frequency is = 0.8111. The legs of the diffraction pattern are as strong as the ordinary B/W aperture, while the resolution is improved by 15.4%

In the cases of the CBQD and CBLD models, the resolution is improved by 20.51% and 15.4% compared to the corresponding resolution in the case of B/W transparent annuli. In addition, the maximum amplitude in the PSF is increased while the total number N is increased. In the CBQD aperture, a value of 224 pixels is shown for N = 10, a value of 230 pixels is computed for N = 20, while in the case of the CBLD aperture, a value of 291 pixels is shown for N = 10.

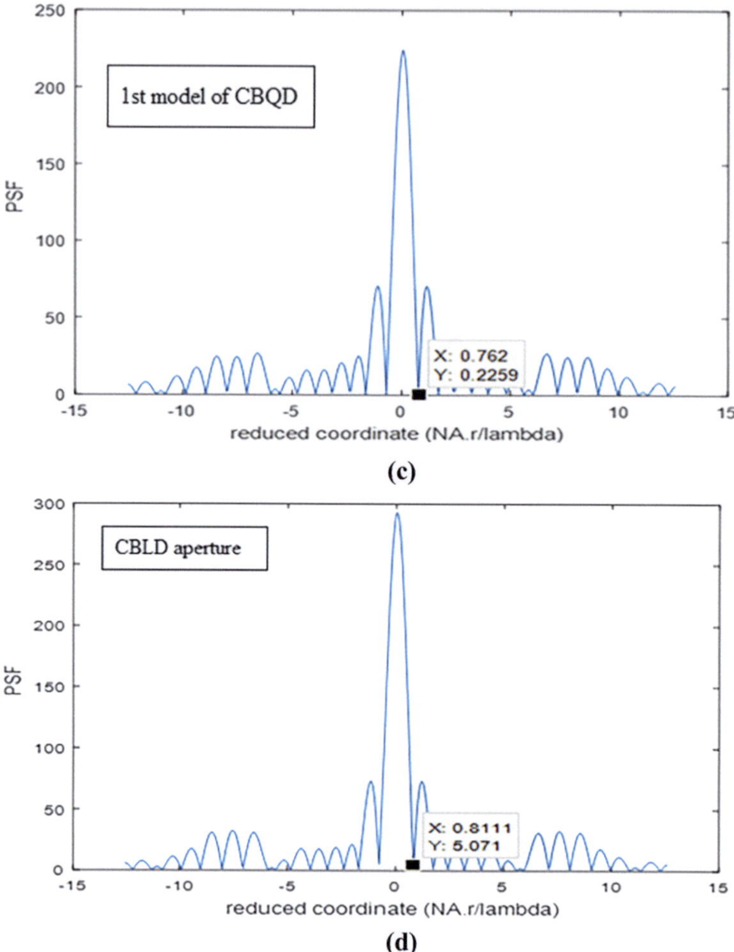

Fig. 4.3 (continued)

A comparison of the PSF for B/W transparent aperture, uniform circular, and annular aperture of a width equal to 5% of the radius, 16 pixels, is shown in Fig. 4.5a, b, and c. The corresponding cut-off spatial frequency in reduced coordinates is respectively $W_c = 0.9586$, 0.9586, and 0.6636. Another comparison for a transparent annulus is represented in Fig. 4.6a, b, and c. The annular widths 0.02, 0.2, and

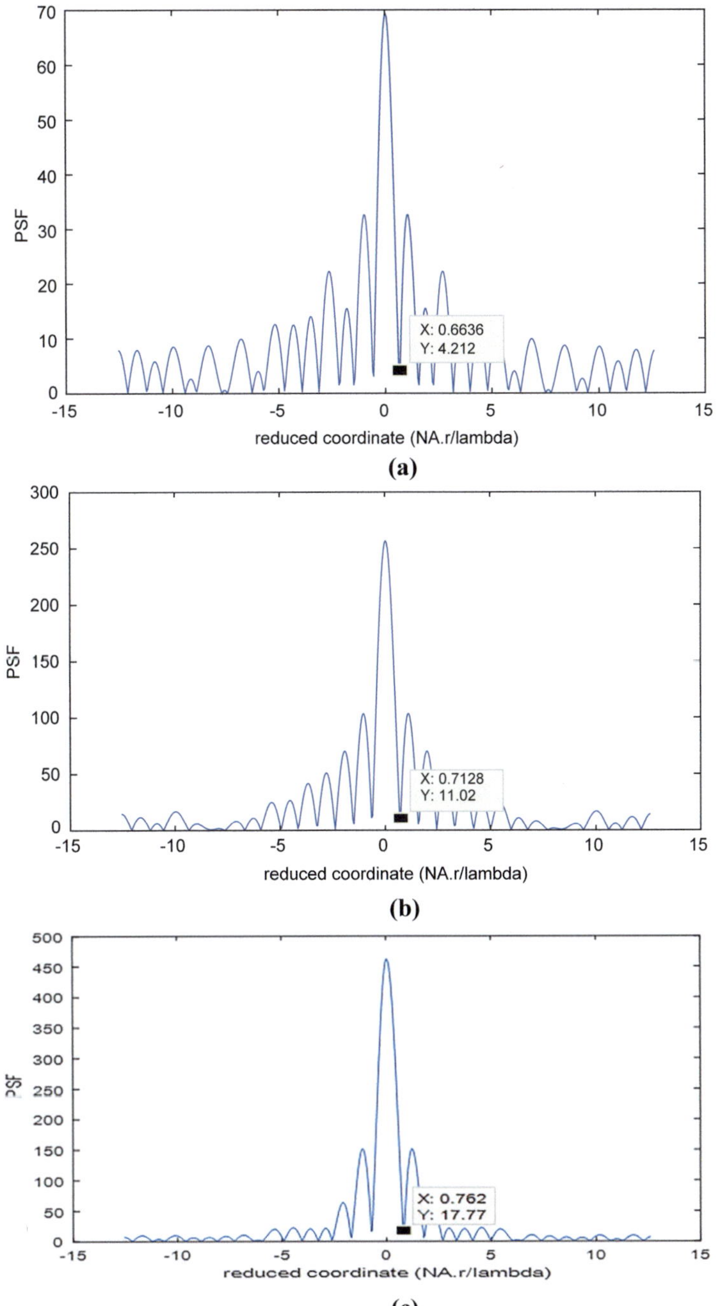

Fig. 4.4   a PSF versus reduced coordinate. Width of the linear annulus = 0.05, external radius = 16 pixels, as shown in Fig. 4.2. **b** PSF versus reduced coordinate. Width of the linear annulus = 0.2, external radius = 16 pixels. **c** PSF versus reduced coordinate. Width of the linear annulus = 0.5, external radius = 16 pixels

Table 4.1 The PSF for different models of apertures and the corresponding cut-off spatial frequency in reduced coordinates. The aperture radius = 16 pixels for all the models

PSF at radius R = 16 pixels	Number of zones (N)	Cut-off spatial frequency $W_c = 2\pi NA.(\frac{r_c}{\lambda})$	Maximum amplitude in pixels
Model 1 (CBQD) ratio 1:1	10	0.762	224
Model 2 (CBQD) ratio 1:1	20	0.762	230
Model 3 (CBQD) ratio 2:1	21	0.762	161
Model 1 (CBLD) ratio 1:1	10	0.8111	291
Model 2 (CBLD) ratio 1:1	20	0.8111	303
Model 3 (CBLD) ratio 2:1	21	0.8111	208
Model 4 linear annulus (width = 0.05)	2	0.6636	70
B/W transparent annuli (CBWD)	10	0.9586	431

0.5 are assumed. In this case, the corresponding cut-off spatial frequency in reduced coordinates is respectively $W_c = 0.6636, 0.6636$, and 0.8111. The external radius is kept constant at 16 pixels. We showed resolution improvement for small annular widths. The legs of the diffraction pattern increased for smaller widths, making it useful for imaging of extended objects.

The reconstructed bone marrow images corresponding to the CBQD aperture (Model 1) are shown in Fig. 4.7a, compared with those obtained using B/W concentric annuli Fig. 4.7b. No difference appeared in the two reconstructed images. The difference is outlined only in the RPSF corresponding to the CBQD and the B/W arrangements. An improvement in resolution in the case of CBQD compared with the case of B/W apertures, since r_c (CBQD) = 0.762$<$ r_c (CBLD) = 0.8111 $<$ r_c (B/W) = 0.9586. Hence, reconstructed images in the case of CBQD are better in resolution compared with B/W concentric annuli for the same maximum radius.

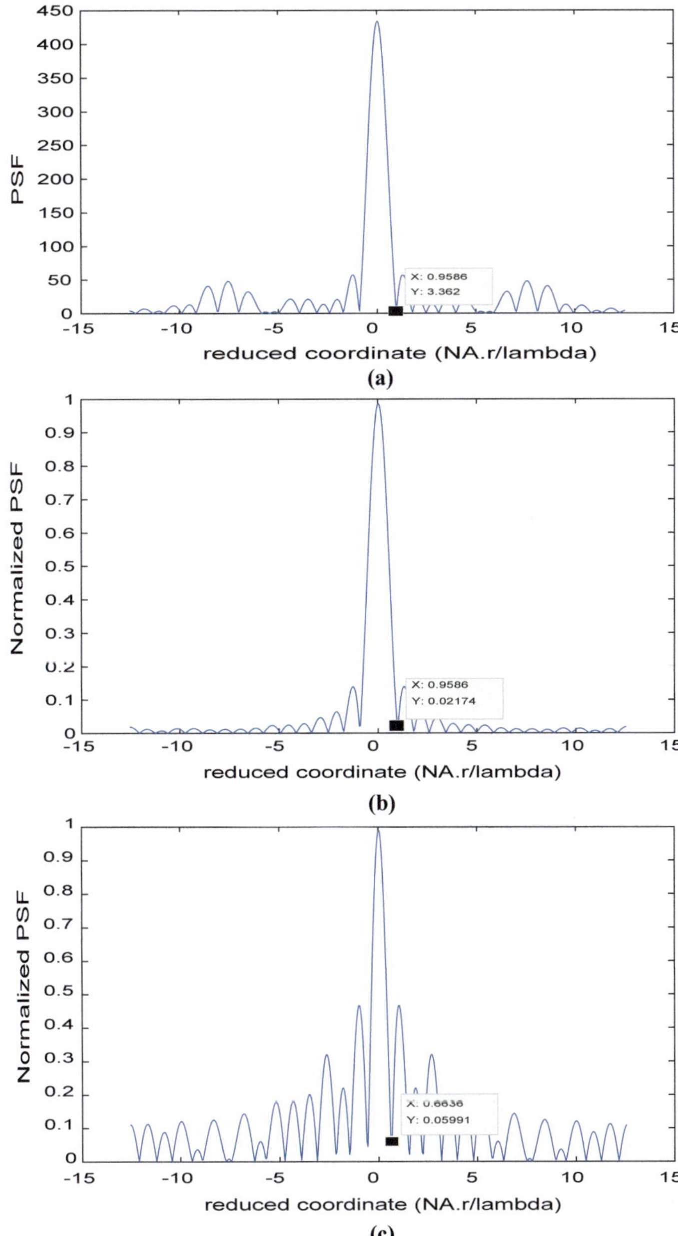

Fig. 4.5 **a** The PSF for B/W transparent aperture where the radius $= 16$ pixels. PSF $_{max.} = 431$ pixels while the cut-off spatial frequency is given as $W_c = 0.9586$. The legs of the diffraction pattern are strong, which is considered useful for imaging extended objects. **b** Normalized PSF for circular uniform aperture where the radius $= 16$ pixels. The cut-off spatial frequency is given as $W_c = 0.9586$. **c** Normalized PSF for annular aperture of width $= 5\%$ from the total radius $= 16$ pixels. The cut-off spatial frequency is found to be $W_c = 0.6636$. The intensity of the diffracting spot is decreased compared with the proposed model

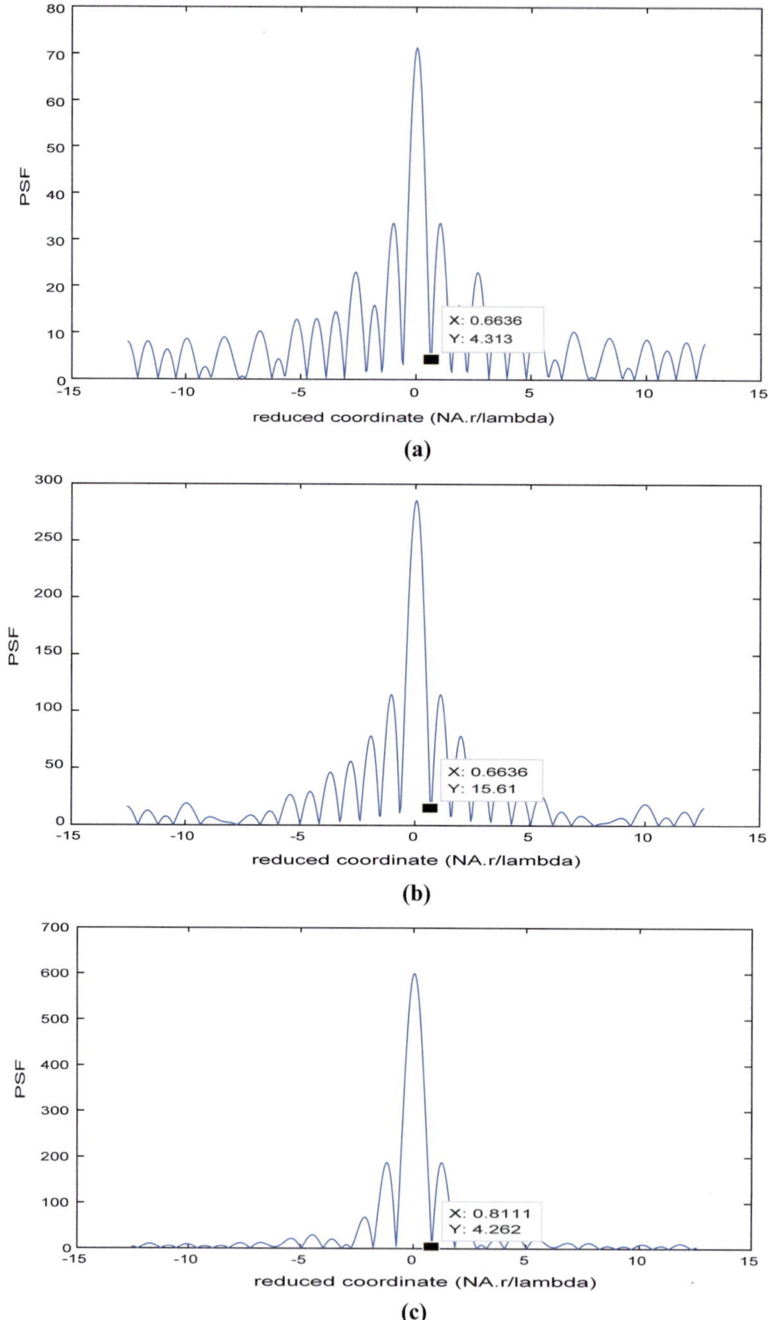

Fig. 4.6 a PSF versus reduced coordinate W. Width of the transparent annulus = 0.05, external radius = 16 pixels. **b** PSF versus reduced coordinate W. Width of the transparent annulus = 0.2, external radius = 16 pixels. **c** PSF versus reduced coordinate W. Width of the transparent annulus = 0.5, external radius = 16 pixels

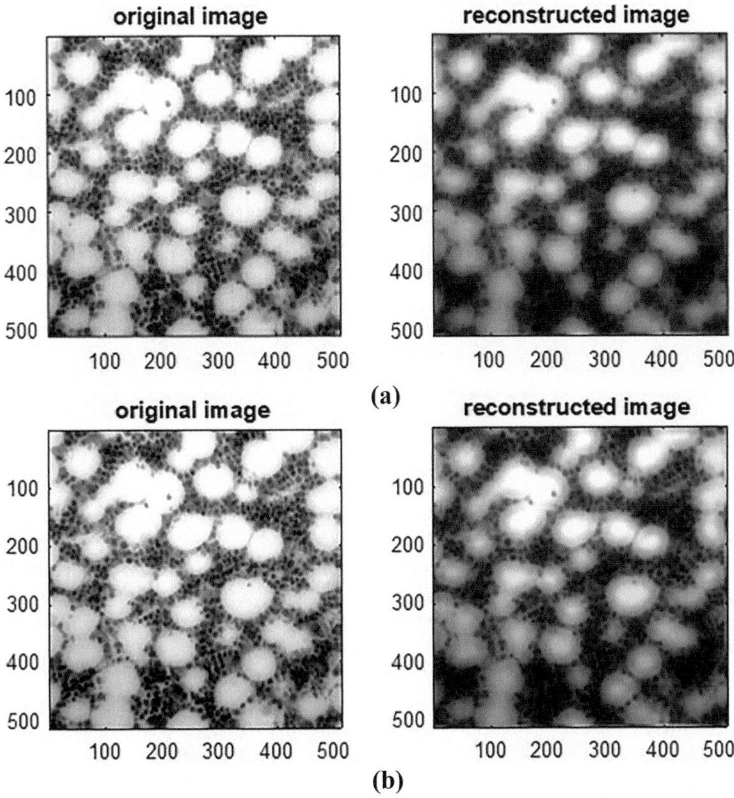

Fig. 4.7 **a** Reconstructed image using the CLSM provided with CBQD apertures for the objective and the collector, N = 10. The original image is of the bone marrow. **b** Reconstructed image using the CLSM provided with B/W apertures for the objective and the collector, N = 10

4.4 Conclusion

The two models of the CBQD apertures showed compromised resolution and contrast compared with circular and annular transparent apertures. The image contrast increases with the number of concentric black and quadratic strips. CBQD has better resolution compared to the ordinary (B/W) transparent annuli and uniform circular apertures; hence, the reconstructed images are better in resolution in the case of CBQD. The suggested CBQD apertures have legs of reasonable amplitude, which are useful for imaging extended objects as compared with the circular uniform aperture.

References

1. M. Minsky, Microscopy apparatus, United States Patent Office. Filed Nov. 7, 1957, granted Dec. 19, 1961. Patent No. 3, 013, 467 (1961)
2. https://bitesizebio.com/19958/what-is-confocal-laser-scanning-microscopy/
3. C.J.R. Sheppard, A. Choudhury, Image Formation in the Scanning Microscope. Opt. Acta **24**, 1051 (1977)
4. C.J.R. Sheppard, Scanned Imagery. J. Phys. D **19**, 2077 (1986)
5. C.J.R. Sheppard, T. Wilson, Image formation in scanning microscopes with partially coherent source and detector. Opt. Acta **25**, 315–325 (1978)
6. C.J.R. Sheppard, D.K. Hamilton, I.J. Cox, Optical microscopy with extended depth of field. Proc. R. Soc., Lond. A, **387**, 171–186 (1983)
7. C.J.R. Sheppard, H.J. Matthews, Imaging in high-aperture optical systems. J. Opt. Soc. Am. **4**, 1354 (1987)
8. H.H. Hopkins, The frequency response of a defocused optical system. Proc. R. Soc., Lond. A, **231**, **91** (1955)
9. L. Levi, R.H. Austing, Determination of Equivalent Pass band of an aberration-free lens using numerical methods. Appl. Optics **7**, 967 (1968)
10. I.J. Cox, C.J.R. Sheppard, Information capacity and resolution in an optical system. J. Opt. Soc. Am. **3**, 1152–1158 (1986)
11. G. Cox, C.J.R. Sheppard, Practical limits of resolution in confocal and non-linear microscopy. Microscopy. Res. Tech. **63**, 18–22 (2004)
12. C.J.R. Sheppard, D.M. Shotton, Image Formation in the Confocal Laser Scanning Microscope. Springer-Verlag, New York Inc.: New York, 15–31 (1997)
13. A.M. Hamed, J.J. Clair, Image and super-resolution in optical coherent microscopes. Optik **64**, 277–284 (1983)
14. A.M. Hamed, J.J. Clair, Studies on optical properties of confocal scanning optical microscope using pupils with radially transmission distribution. Optik **65**, 209–218 (1983)
15. A.M. Hamed, Improvement of point spread function (PSF) using linear-quadratic aperture. Optik **131**, 838–849 (2017)
16. C.J.R. Sheppard, X.Q. Ma, Confocal microscopes with slit apertures. J. Modern Optics **35**, 1169–1185 (1988)
17. A.M. Hamed, Design of a cascaded black-linear distribution (CBLD) in circular aperture and its application on confocal laser scanning microscope (CLSM). Am. J. Opt. Photonics **7**(3), 46–56 (2019)

Chapter 5
Design of an Improved Aperture and an Application in Microscopy

5.1 Introduction

Many authors [1–5] investigated annular apertures. The point spread function (PSF) has high sidelobes compared to uniform circular apertures. However, these annular apertures improve the resolution while decreasing the image contrast. The use of lenses with annular apertures was presented [6, 7].

Intensive studies on aperture modulation are presented to improve microscope resolution and image contrast [8–10]. Apertures of linear and quadratic distributions [8, 11] were studied using a confocal scanning laser microscope. We considered the investigation of the PSF to be an indication of the resolution of the microscope.

Recently, many applications have carried out speckle imaging using modulated apertures and interferometry [12–14], leading to the recognition of the apertures. We have improved the contrast of speckle images in [15, 16]. We computed the surface roughness of objects using optical correlation [17].

Other articles are provided on aperture modulation to increase microscope resolution [18–20]. We obtained further improvement in resolution for a confocal laser scanning microscope (CLSM), where the PSF is computed from the product of the PSF corresponding to each aperture [21, 22]. Recently, we considered a new B/W concentric hexagonal aperture, and we got the PSF and discussed the resolution limit computed from the cut-off spatial frequency in the diffraction plane [23]. The annular Hermite–Gaussian aperture is considered to form speckle images using a laser beam [23, 24]. Other authors in [25–27] presented a work to improve spatial resolution using focal modulation microscopy.

In this chapter, we suggest two modulated apertures. The first aperture has a distribution oscillating between linear and conic (L/C) to improve the PSF corresponding to the aperture. The second is a hexagonal annulus where the central disc has a transparent linear distribution. A theoretical analysis is presented, followed by results and discussion, and finally, a conclusion is given.

A. M. Hamed, *Image Processing Techniques for Deformed and Sparse Aperture Systems*,
SpringerBriefs in Applied Sciences and Technology,
https://doi.org/10.1007/978-3-032-04921-6_5

5.2   Theoretical Analysis

5.2.1   The First Model of Linear-Conic Configuration

We assume that the aperture of a cascaded linear-conic distribution (L/C) is represented as follows:

$$P_{L-C}(\rho) = P_{1L}(\rho) + P_{2C}(\rho) + P_{3L}(\rho) + P_{4C}(\rho) + P_{5L}(\rho) + P_{6C}(\rho) + P_{7L}(\rho) + P_{8C}(\rho)$$

$$(5.1)$$

In the aperture represented by Eq. (5.1), we considered a cascaded (L/C) of eight-zones, with

$$P_{1L}(\rho) = \rho; \text{ for } 0 \leq \left| \frac{\rho}{\rho_0} \right| \leq 0.125,$$

$$P_{2C}(\rho) = 1 - \rho; \text{ for } 0.125 \leq \left| \frac{\rho}{\rho_0} \right| \leq 0.25,$$

$$P_{3L}(\rho) = \rho; \text{ for } 0.25 \leq \left| \frac{\rho}{\rho_0} \right| \leq 0.375,$$

$$P_{4C}(\rho) = 1 - \rho; \text{ for } 0.375 \leq \left| \frac{\rho}{\rho_0} \right| \leq 0.5,$$

$$P_{5L}(\rho) = \rho; \text{ for } 0.5 \leq \left| \frac{\rho}{\rho_0} \right| \leq 0.625,$$

$$P_{6C}(\rho) = 1 - \rho; \text{ for } 0.625 \leq \left| \frac{\rho}{\rho_0} \right| \leq 0.75,$$

$$P_{7L}(\rho) = \rho; \text{ for } 0.75 \leq \left| \frac{\rho}{\rho_0} \right| \leq 0.875,$$

$$P_{8C}(\rho) = 1 - \rho; \text{ for } 0.875 \left| \frac{\rho}{\rho_0} \right| \leq 1.0$$

$$(5.2)$$

We observe that the linear distribution is the inverse of the conic distribution, where the central zone has a linear distribution.

We plotted the image corresponding to the cascaded (L/C) aperture represented by Eqs. (5.1), (5.2) as shown in Fig. 5.1.

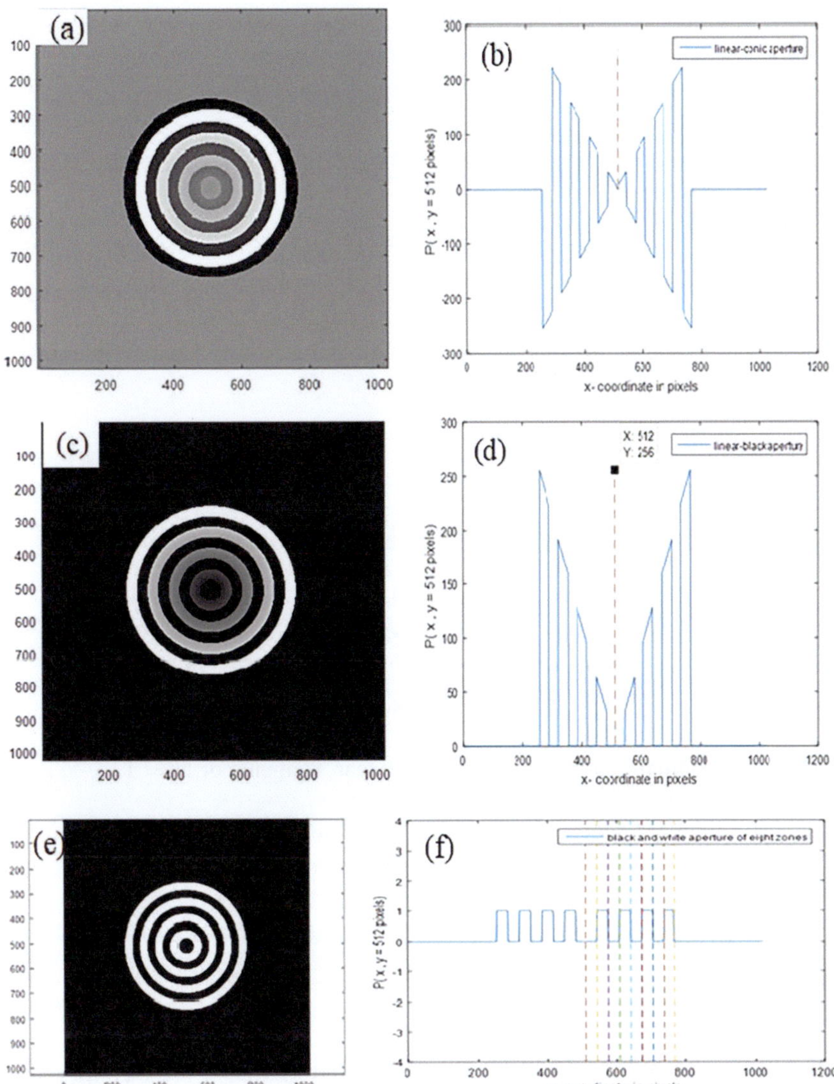

Fig. 5.1 Images of the linear-conic (L/C), linear-black (L/B), and black, white (B/W) apertures in a matrix of dimensions 1024 × 1024 pixels and their plots at y = 512 pixels are shown. The aperture radius = 256 pixels, and the zone widths are 32 pixels

Now, we compute the PSF corresponding to the described aperture as follows [8].

For example, we solve Eq. (5.2) for the linear segment 7, by applying the two-dimensional Fourier transform in radial coordinates as follows:

$$h_{7L}(r) = \int_0^{2\pi} \int_{0.75\rho_0}^{0.875\rho_0} P_{7L}(\rho) \exp\{-j(2\pi/\lambda f)\rho r \cos(\theta - \phi)\}\rho \, d\rho \, d\theta \qquad (5.3)$$

$$= 2\pi \int_{0.75\rho_0}^{0.875\rho_0} \rho^2 J_0\left[\left(\frac{2\pi}{\lambda f}\right)\rho r\right] d\rho \qquad (5.4)$$

To solve Eq. (5.4), we divided the definite integral into two separate integrals as follows:

$$h_{7L}(r) = 2\pi \left\{ \int_0^{0.875\rho_0} \rho^2 J_0\left[\left(\frac{2\pi}{\lambda f}\right)\rho r\right] d\rho - \int_0^{0.75\rho_0} \rho^2 J_0\left[\left(\frac{2\pi}{\lambda f}\right)\rho r\right] d\rho \right\} \qquad (5.5)$$

For this segment, we finally obtained the following result [8]:

$$h_{7L}(w) = 2\pi \left\{ \frac{J_1(o.875w)}{(0.875w)} + \frac{J_0(0.875w)}{(0.875w)^2} - 2\frac{\sum_i J_i(o.875w)}{(0.875w)^2} \right\}$$
$$- 2\pi \left\{ \frac{J_1(o.75w)}{(0.75w)} + \frac{J_0(0.75w)}{(0.75w)^2} - 2\frac{\sum_i J_i(o.75w)}{(0.75w)^2} \right\} \qquad (5.6)$$

Hence, we obtain similar results for the linear segments of $N = 1, 3, 5,$ and 7 corresponding to the defined zones in the aperture.

We solved the last segment in Eq. (5.2) in the aperture in the radial range from $0.875 \, \rho_0$ up to ρ_0 to obtain the following result corresponding to the last conic zone:

$$h_{8C}(w) = 2\pi \left\{ \frac{[2\sum_i J_i(w) - wJ_0(w)]}{w^3} \right\} - 2\pi \left\{ \frac{[2\sum_i J_i(0.875w) - 0.875wJ_0(0.875w)]}{(o.875w)^3} \right\} \qquad (5.7)$$

In Eqs. (5.6), (5.7), $w = \left[\left(\frac{2\pi}{\lambda f}\right)\rho_0 r\right]$ is the reduced coordinate in the Fourier plane of radial coordinate r.

By grouping the PSF corresponding to all zones, we obtain the PSF corresponding to the cascaded linear-conic aperture.

5.2.2 Case of Two Concentric Layers (Triangular Distribution)

To confirm the FFT used to obtain the PSF in Eq. (5.2), we consider only two concentric layers of the triangular distribution written as follows:

$$P_{L-C}(\rho) = \rho; \text{ for } 0 \leq |\frac{\rho}{\rho_0}| \leq 0.5, = 1 - \rho; \text{ for } 0.5 \leq |\frac{\rho}{\rho_0}| \leq 1 \tag{5.8}$$

By using Eqs. (5.6) and (5.7), we obtain the PSF for the two layers, one linear and the other conic, as follows:

$$h(w) = 2\pi \left\{ \frac{J_1(o.5w)}{(0.5w)} + \frac{J_0(0.5w)}{(0.5w)^2} - 2\frac{\sum_i J_i(o.5w)}{(0.5w)^2} \right\}$$
$$+ 2\pi \left\{ \frac{[2\sum_i J_i(w) - wJ_0(w)]}{w^3} \right\} - 2\pi \left\{ \frac{[2\sum_i J_i(0.5w) - 0.5wJ_0(0.5w)]}{(o.5w)^3} \right\}$$
$$\tag{5.9}$$

5.2.3 The Second Model of a Hexagonal Annulus with a Central Disc of a Linear Distribution

We summarized the PSF results for the hexagonal aperture recently presented [21] as follows, where the aperture is shown in Fig. 5.8:

$$P(x, y) = \text{rect}(x, y) + \text{tri}\left(x, y - \frac{2a}{3}\right) + \text{tri}(x, y + \frac{2a}{3}) \tag{5.10}$$

where $\text{rect}(x, y) = 1; |\frac{x}{b}| \leq 1$ and $|\frac{y}{b}| \leq 1$. It represents a rectangle of sides a and b where $a = b$.

$$\text{tri}\left(x, y - \frac{2a}{3}\right) = 1.$$ It represents a triangle of base 2b and height a.

We operated the Fourier transform using Eq. (5.10), and we got the PSF (Hamed 2023) as follows:

$$h(u, v) = \left[\frac{\sin(\pi bu)}{\pi bu}\right]\left[\frac{\sin(\pi av)}{\pi av}\right]\left\{1 + 2\left[\frac{\sin(\pi bu)}{\pi bu}\right]\left[\frac{\sin(\pi av)}{\pi av}\right]\cos\left[\left(\frac{4\pi a}{3\lambda f}\right)v\right]\right\}$$

$$(5.11)$$

In this study, we represent the hexagonal annulus as the difference between two hexagons, and add a disc in the center with a linearly distributed function, as shown in Fig. 5.9. The aperture is represented mathematically as follows:

$$P(x, y) = \left\{\text{rect}(x, y) + \text{tri}\left(x, y - \frac{2a}{3}\right) + \text{tri}\left(x, y + \frac{2a}{3}\right)\right\}$$

$$- \left\{\text{rect}(\alpha x, \ \beta y) + \text{tri}\left(\alpha x, \beta y - \frac{2a}{3}\right) + \text{tri}\left(\alpha x, \beta y + \frac{2a}{3}\right)\right\} + P_{linear}(x, y)$$

$$(5.12)$$

where $\alpha = 0.9$, $\alpha = \beta$; hence the width of the annulus $= 0.1$.

The central disc is described by a linear distribution with a radius $= a/2$, and is represented as follows:

$$P_{\text{linear}}(\rho) = \rho \text{ for} \left|\frac{\rho}{\rho_0}\right| \leq 1, = 0 \text{ for} \left|\frac{\rho}{\rho_0}\right| \geq 1 \qquad (5.13)$$

where $\rho = \sqrt{x^2 + y^2}$. The radial coordinate corresponds to the Cartesian coordinates (x, y) in the plane of the aperture.

We apply the Fourier transform operation to the aperture described in Eq. (5.12) to obtain the PSF as follows:

$$h(r) = \text{F.T.}\left\{\text{rect}(x, y) + \text{tri}\left(x, y - \frac{2a}{3}\right) + \text{tri}\left(x, y + \frac{2a}{3}\right)\right\}$$

$$-\text{F.T.}\left\{\text{rect}(\alpha x, \beta y) + \text{tri}\left(\alpha x, y - \beta\frac{2a}{3}\right) + \text{tri}\left(\alpha x, y + \beta\frac{2a}{3}\right)\right\} \qquad (5.14)$$

$$+\text{F.T.}\{P_{\text{linear}}(\rho)\}$$

For $\alpha = 0.9$, $\alpha = \beta$. Equation (5.14) is rewritten as follows:

$$h(r) = \text{F.T.}\left\{\text{rect}(x, y) + \text{tri}\left(x, y - \frac{2a}{3}\right) + \text{tri}\left(x, y + \frac{2a}{3}\right)\right\}$$

$$-\text{F.T.}\left\{\text{rect}\left(\frac{4}{5}x, \frac{4}{5}y\right) + \text{tri}\left(\frac{4}{5}x, y - \frac{3a}{5}\right) + \text{tri}\left(\frac{4}{5}x, y + \frac{3a}{5}\right)\right\} \qquad (5.15)$$

$$+\text{F.T.}\{P_{\text{linear}}(x, y)\}$$

Finally, after operating the transformations in Eq. (5.15), we obtain the following result, considering that the central disc has a linear distribution [8, 11]:

$$
\begin{aligned}
h(r) = {} & \left[\frac{\sin(\pi bu)}{\pi bu}\right]\left[\frac{\sin(\pi av)}{\pi av}\right]\left\{1 + 2\left[\frac{\sin(\pi bu)}{\pi bu}\right]\left[\frac{\sin(\pi av)}{\pi av}\right]\cos\left(\frac{4\pi a}{3\lambda f}\right)v\right\} \\
& -\left[\frac{\sin(0.9\pi bu)}{(0.9\pi bu)}\right]\left[\frac{\sin(0.9\pi av)}{(0.9\pi av)}\right]\left\{1 + 2\left[\frac{\sin(0.9\pi bu)}{(0.9\pi bu)}\right]\left[\frac{\sin(0.9\pi av)}{(0.9\pi av)}\right]\cos\left(\frac{6\pi a}{5\lambda f}\right)v\right\} \\
& + 2\left[\frac{J_1(w)}{w} + \frac{J_0(w)}{w^2} - 2\sum_i \frac{J_i(w)}{w^3}\right]; i = 1, 3, 5, \ldots
\end{aligned}
$$

$$(5.16)$$

where $r = \sqrt{u^2 + v^2}$ is the radial coordinate in the Fourier plane (u, v) and $w = \pi u\sqrt{2}$ is the reduced coordinate in the Fourier plane.

5.2.4 Intensity Distribution in CSLM

The intensity distribution, using the confocal scanning laser microscope (CSLM), is computed using the following formula [7, 8]:

$$I(r) = |\ g(r) \otimes h_T(r)\ |^2 \tag{5.17}$$

g(r) represents the complex amplitude of the image in the detection plane of coordinates (x, y).

$h_T(r)$ is the resultant point spread function (RPSF) and is equal to the product of the PSF corresponding to the illumination objective and the PSF corresponding to the detection objective, i.e., $h_T(r) = h_1(r).h_2(r)$. In this CSLM, the Fourier transform of the total PSF is the coherent transfer function (CTF), i.e.,

$$CTF(\rho) = F.T.\{h_1(r).h_2(r)\} = P_1(\rho) \otimes P_2(\rho) \tag{5.18}$$

Hence, CTF is the convolution product corresponding to apertures P_1 and P_1 of the objectives L_1 and L_2.

5.3   Results and Discussion

We plotted the image corresponding to the cascaded linear-conic (L/C) aperture using MATLAB code as shown in Fig. 5.1a. A line plot at $y = 512$ pixels corresponding to the (L/C) image is shown in Fig. 5.1b. We started with a linear distribution from the center, followed by a conic distribution. We considered that the total number of linear zones equals the total number of conic zones $= 4$. In the figure, the width of each zone is equal to 32 pixels, and for the total number of zones, we have the maximum radius $= 256$ pixels. In addition, we plotted the aperture with a concentric linear-black (L/B) distribution in Fig. 5.1c and its line plot at $y = 512$ pixels, shown in Fig. 5.1d. The image of the black and white (B/W) aperture is shown in Fig. 5.1e. The corresponding plot at $y = 512$ pixels is shown in Fig. 5.1f.

We computed the PSF by applying the FFT to the fabricated apertures. The PSF corresponding to the (L/C) aperture is shown in Fig. 5.2a. Referring to the plot, the cut-off spatial frequency $= 66$ pixels. We observed six strong sidelobes on both sides around the central peak.

We computed the PSF corresponding to the L/B aperture and plotted it in Fig. 5.2b. The cut-off spatial frequency $= 69$ pixels. We observed a central peak followed

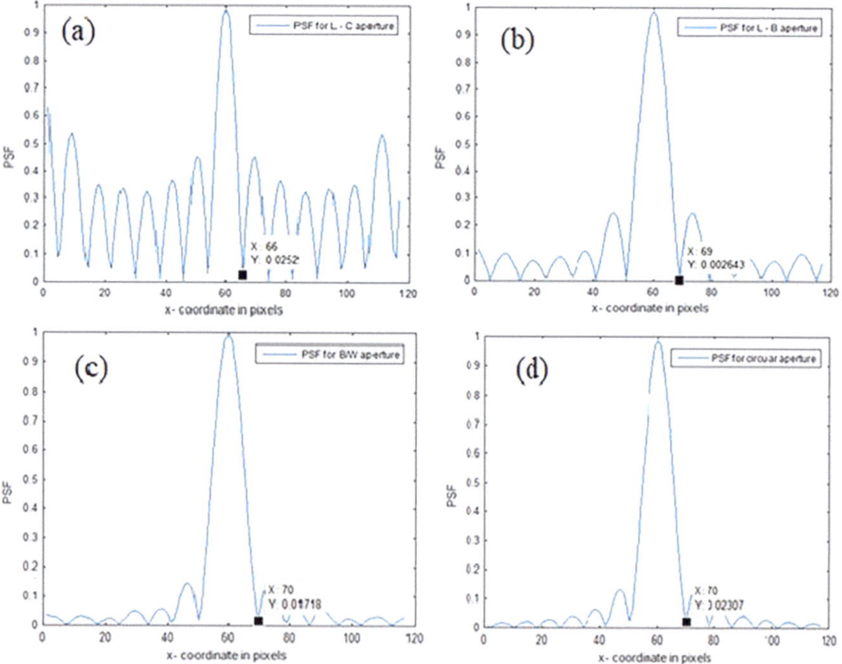

Fig. 5.2 PSF corresponding to the linear-conic (L/C), linear-black (L/B), and black-and-white (B/W) apertures compared to the PSF corresponding to the transparent circular aperture. The cut-off spatial frequency is shown in the plot

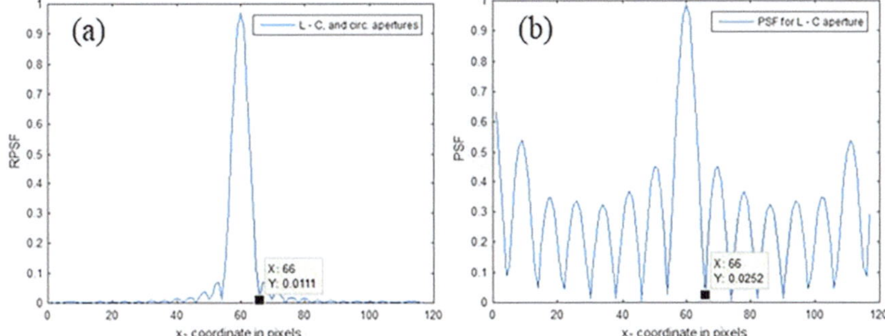

Fig. 5.3 RPSF corresponds to two different apertures where the illuminating aperture is uniform circular and the other has 8 concentric zones of (L/C) distribution. The cut-off equals 66 pixels, like the cut-off obtained for the aperture (L/C) alone. The interest is the attenuation of the secondary peaks. Hence, this arrangement is useful in confocal scanning laser microscopes

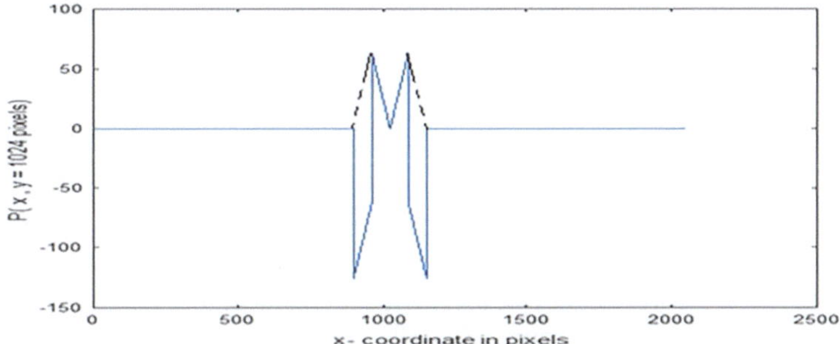

Fig. 5.4 Plot of two concentric linear and conic zones. The aperture radius equals 128 pixels, which is equal to the sum of the two equal triangles, each with a base = 64 pixels

by sidelobes of moderate intensities compared to the results in Fig. 5.2a. The PSF computed for the B/W aperture has a cut-off spatial frequency = 70 pixels as shown in Fig. 5.2c. The PSF corresponding to a uniform circular aperture is plotted in Fig. 5.2d for comparison with the (L/C) and (L/B) models. The cut-off spatial frequency = 70 pixels for the transparent circular aperture.

The RPSF corresponds to two different apertures, where the illuminating aperture is uniform circular and the other for detection has an (L/C) distribution as shown in Fig. 5.3a and compared to the PSF for the (L/C) aperture alone as shown in Fig. 5.3b. The legs in the RPSF Fig. 5.3a are diminished compared to the results of PSF for (L/C) aperture in Fig. 5.2a.

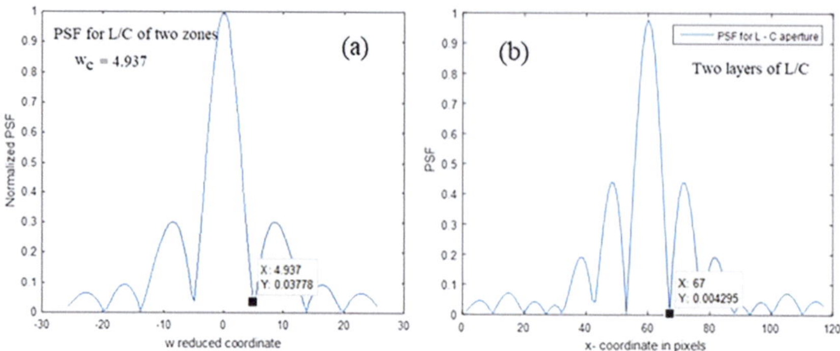

Fig. 5.5 The theoretical curve for PSF, using the aperture plotted in Fig. 5.4, is represented as shown in Fig. 5.5a. The PSF plot using FFT is given for comparison, in Fig. 5.5b. The external radius = 128 pixels, and the matrix of dimensions 2048 × 2048 pixels

We plotted an aperture composed of two concentric linear and conic zones to facilitate the computation of the theoretical PSF curve. The aperture radius = 128 pixels is equivalent to the sum of the two equal triangles of base = 64 pixels, as shown in Fig. 5.4.

The theoretical curve for PSF computed from Eq. (5.9), using the aperture plotted in Fig. 5.4, is represented as shown in Fig. 5.5 a. The PSF plot using FFT is given for comparison as in Fig. 5.5b. The external radius = 128 pixels, and the matrix dimensions are 2048 × 2048 pixels.

We showed the PSF theoretical plots corresponding to the different apertures, linear, conic, linear-conic, and circular, in Fig. 5.6a and b.

Referring to Fig. 5.6a, the resolution in the former case of L/C aperture lies between the separate linear and conic resolutions. The cut-off spatial frequency in reduced coordinates obtained from the PSF plots follows this inequality:

$$w_c \ (linear) = 3.378 < w_c(L/C) = 4.937 < w_c \ (conic) = 5.976$$

Referring to Fig. 5.6a and b, we showed that the lower cut-off is attained in linear apertures Fig. 5.6b; hence we obtain better resolution compared to the results of a transparent circular aperture, since the cut-off spatial frequency obtained from the PSF plots has the following:

$$w_c \ (linear) = 3.378 < w_c \ (circular) = 3.897$$

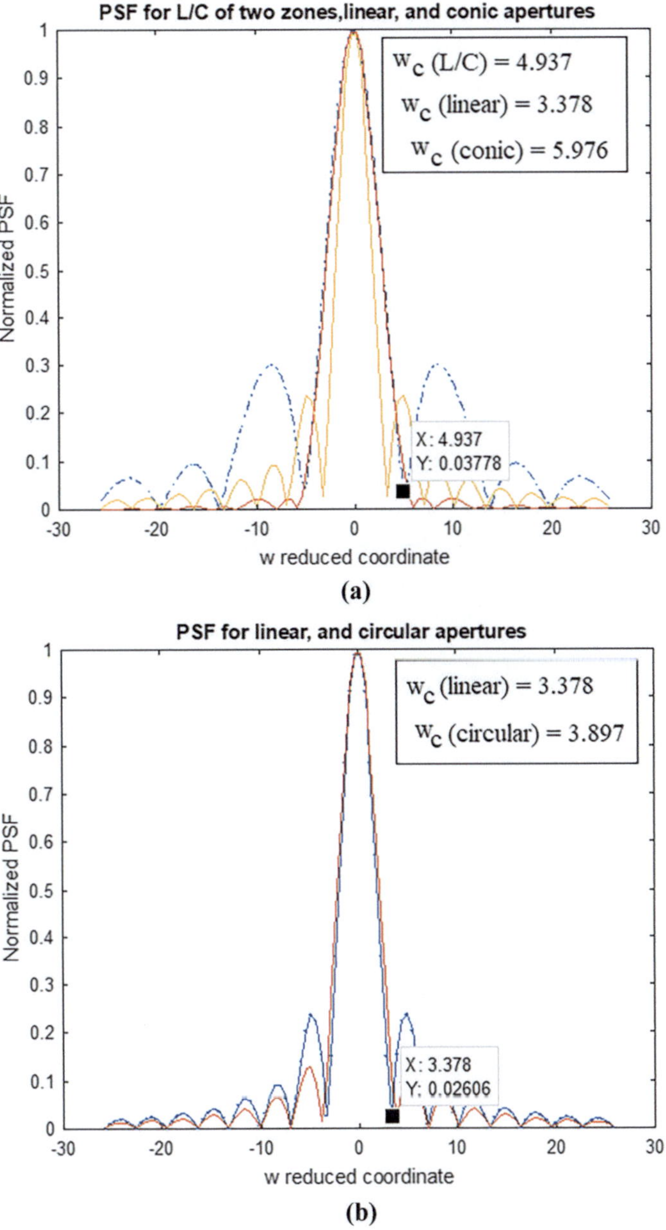

Fig. 5.6 **a** The theoretical curve for the PSF is plotted for the L/C aperture and compared with the PSF corresponding to the linear and conic apertures. The cut-off spatial frequency w_c is shown **b** The PSF corresponding to the linear aperture and compared with the PSF for a transparent circular aperture

The reconstructed image using CSLM, provided with two identical apertures, each with (L/C) distributions, is plotted as in Fig. 5.7a. A reconstructed image with two different apertures, one of (L/C) distribution and the other of constant distribution, using the CSLM is shown in Fig. 5.7b.

For the second model, we computed the PSF for the different modulated hexagonal apertures and compared it with that corresponding to the transparent hexagonal aperture using the analytical formula 5.16. We obtained different cut-offs depending on the width of the hexagonal annulus and the disc with linear distribution.

The aperture used in the processing is shown in Fig. 5.9. It has a transparent hexagonal annulus with a central linearly distributed disc.

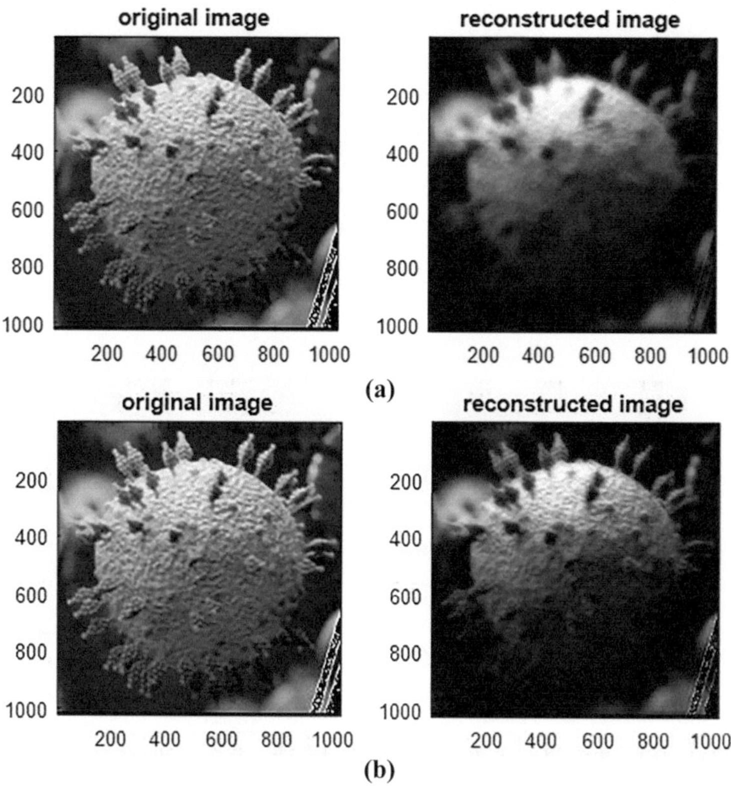

Fig. 5.7 **a** Reconstructed image using CSLM with two identical apertures, each of eight concentric zones of linear-conic distribution. **b** Reconstructed image using CSLM with two different apertures, one of eight concentric zones of linear-conic distribution, and the other of constant distribution using transparent circular apertures

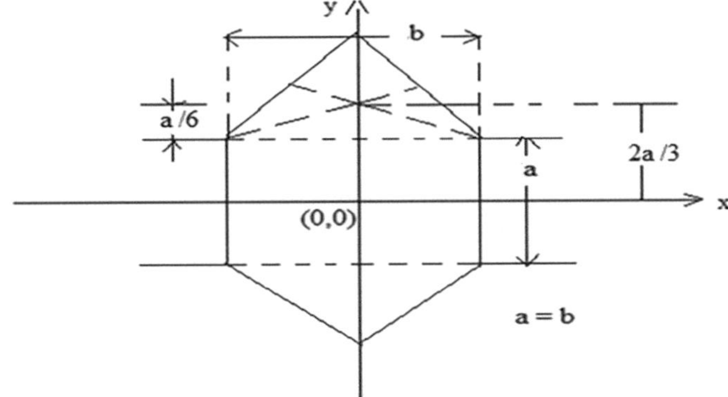

Fig. 5.8 Shape of the hexagonal aperture. The side length is equal to a. The distance from the triangle center to the hexagonal center, a/2 + a/6, is 2a/3

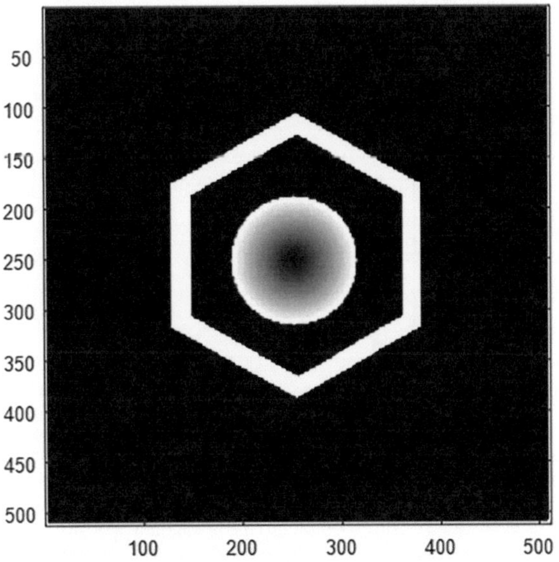

Fig. 5.9 The aperture used in the processing. It has a transparent hexagonal annulus with a central disc with a linear distribution

We plotted the PSF, using Eq. 5.16 except the linear term, corresponding to the transparent hexagonal aperture with $a = b = 1$, shown in Fig. 5.10a. The PSF cut-off = 0.6347. The PSF corresponding to the hexagonal annulus is shown in Fig. 5.10b, and the cut-off = 0.3808, where the annular width is 0.2. Another PSF curve corresponds to the hexagonal annulus where the annular width of 0.5 is shown in Fig. 5.10c, and the corresponding cut-off is 0.4443. The range x is from 0 to 2π for all the plots shown in Fig. 5.10.

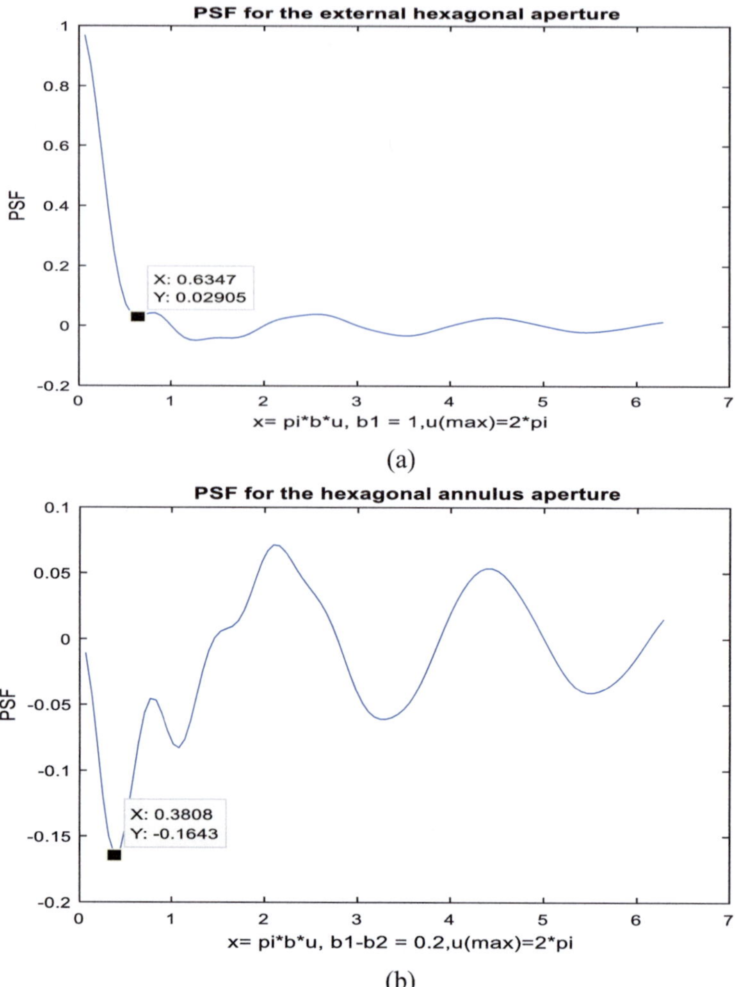

Fig. 5.10 **a** PSF corresponding to the hexagonal aperture of $a = b = 1$, and the range x is from 0 to 2π. As shown in Fig. 5.10 a, the cut-off $= 0.6347$. **b** PSF corresponding to the hexagonal annulus computed from the difference between the external and internal hexagonal apertures, and range x is from 0 to 2π. The cut-off $= 0.3808$, where the annular width is 0.2 as shown in Fig. 5.10b. **c** PSF corresponding to the hexagonal annulus computed from the difference between the external and internal hexagonal apertures, and the range of x is extended from 0 up to 2π. As shown in Fig. 5.10c, the cut-off $= 0.4443$, where the annular width is 0.5

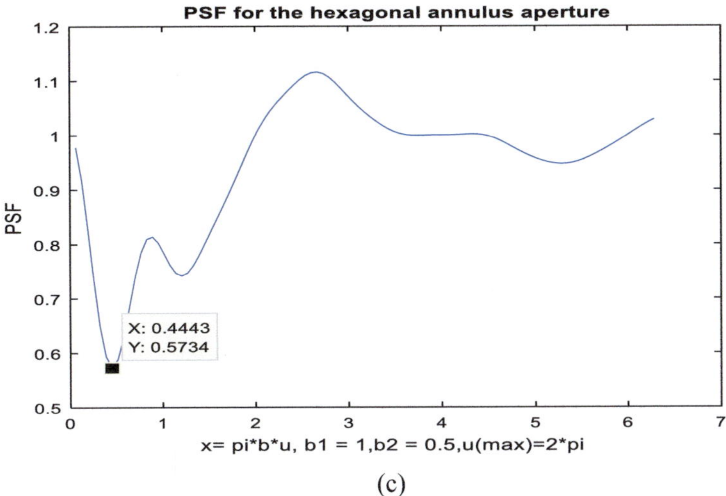

(c)

Fig. 5.10 (continued)

Referring to Fig. 5.10b, c, an improved resolution is shown. It is obtained by narrowing the aperture width as expected. In addition, the resolution for the hexagonal annulus is better than the transparent hexagonal aperture, as in Fig. 5.10a. In addition, referring to Fig. 5.10b, c, the rings are strengthened outside the central band for the hexagonal annular arrangement compared with the weak harmonics shown in Fig. 5.10a.

We computed the theoretical PSF corresponding to the annular hexagonal aperture combined with the central linear disc, using Eq. (5.16). We represented the PSF curve in Fig. 11a for an annular width of 0.2. The corresponding cut-off = 0.6029. Another PSF curve is shown in Fig. 5.11 b for an annular width of 0.5. The cut-off = 0.6347.

Referring to Fig. 5.11a and b, we obtained further improvement in resolution for a smaller annular width, giving a small cut-off value as expected.

In addition, by comparing Fig. 5.11a and b for the annular hexagonal aperture combined with the central linear disc with Fig. 5.10a–c corresponding to the transparent hexagonal aperture, and the hexagonal annulus, we showed that the optimum aperture for the resolution improvement is the annular hexagonal aperture. The transparent hexagonal aperture is poorer in resolution than the hexagonal annulus and the annular hexagonal aperture combined with the central linear disc. In addition, the aperture under investigation has a resolution value between the resolution of the annular hexagonal and the transparent hexagonal apertures. The optimum contrast is unity for the transparent hexagonal aperture. The contrast is improved for the aperture

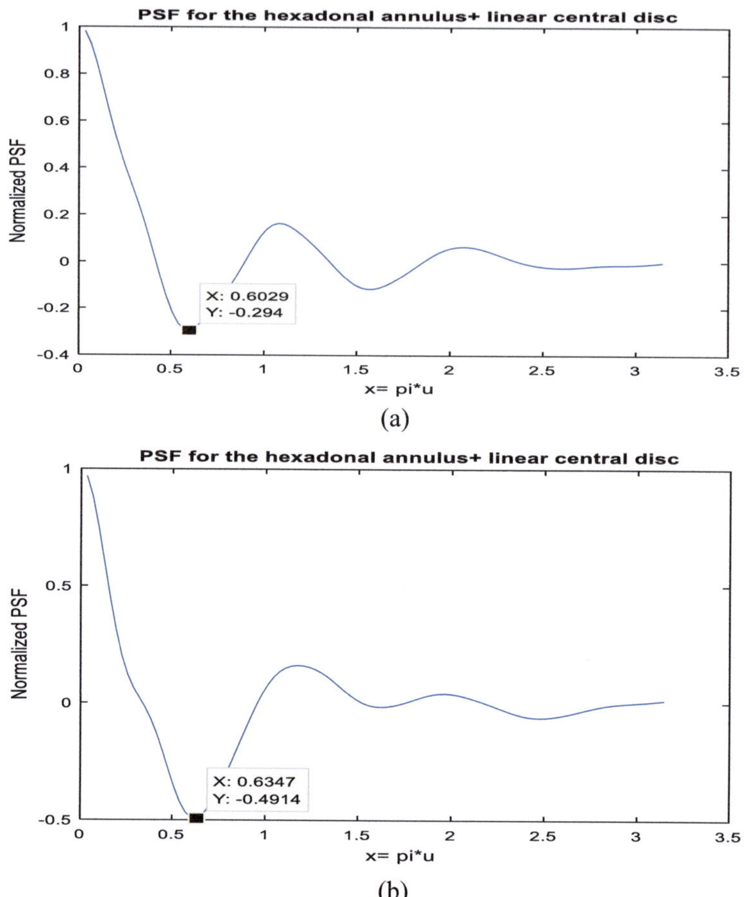

Fig. 5.11 **a** PSF corresponding to the hexagonal annulus combined with a central disc of linear distribution as shown in Fig. 5.9. The range of x is from 0 to π. The cut-off $= 0.6029$, where the annular width is 0.2 **b** PSF corresponding to the hexagonal annulus combined with a central disc of linear distribution as shown in Fig. 5.9. The range of x is extended from 0 up to π. The cut-off $= 0.6347$, where the annular width is 0.5

under investigation compared to the pure hexagonal annulus. Hence, compromising resolution and contrast are realized using the aperture under investigation. These apertures are useful for imaging using confocal microscopy since the detection has a very weak intensity.

We computed the PSF by applying the Fast Fourier transform, which corresponds to the transparent hexagonal aperture. It is represented in Fig. 5.12a. The corresponding total bandwidth is $=8$ pixels. The PSF corresponding to the aperture of a hexagonal annulus combined with a central linear distribution is shown in Fig. 5.12b. The PSF corresponding to the aperture of the hexagonal annulus is shown in Fig. 5.12c. It is shown that the total bandwidth equals 4 pixels in Fig. 5.12b and c, compared with 8 pixels obtained from Fig. 5.12a for the transparent hexagonal aperture. Hence, the resolution is improved in the case of the hexagonal annulus and the aperture under investigation. All apertures shown in Fig. 5.12a–c have dimensions 512×512 pixels.

We tabulated the results obtained from the PSF cut-off corresponding to each modulated aperture as shown in Table 5.1.

We plotted the autocorrelation corresponding to the hexagonal annulus combined with a central disc of linear distribution compared with that corresponding to the transparent hexagonal and hexagonal annular apertures, as in Fig. 5.13. The autocorrelation bandwidth is two times the maximum diameter of the apertures. The autocorrelation or the CTF corresponding to the Coronavirus image is plotted in Fig. 5.14.

We plotted in Fig. 5.15a the reconstructed image of the Coronavirus using the CSLM with the two objectives of the microscope, hexagonal annuli combined with a central disc of linear distribution. The reconstructed images using the CSLM provided with objectives with uniform hexagonal and hexagonal annulus are given for comparison as in Fig. 5.15b, c. The aperture widths in all the plots are equal to 256 pixels. Referring to Fig. 5.12b and c, the cut-off in the PSF is the same and equal to 4 pixels, and the resolution in the reconstructed image using either the hexagonal annulus provided with the central disc of a linear distribution, or the hexagonal annular aperture is invariant. The contrast is improved in the case of the hexagonal annulus with the central disc of linear distribution.

5.4 Conclusion

We deduce that the PSF corresponding to the linear-conic (L/C) aperture has many advantages compared to the linear-black (L/B) and black-and-white (B/W) apertures.

First, the (L/C) aperture has many peaks outside the center of the PSF pattern compared to the (L/B) and (B/W) apertures.

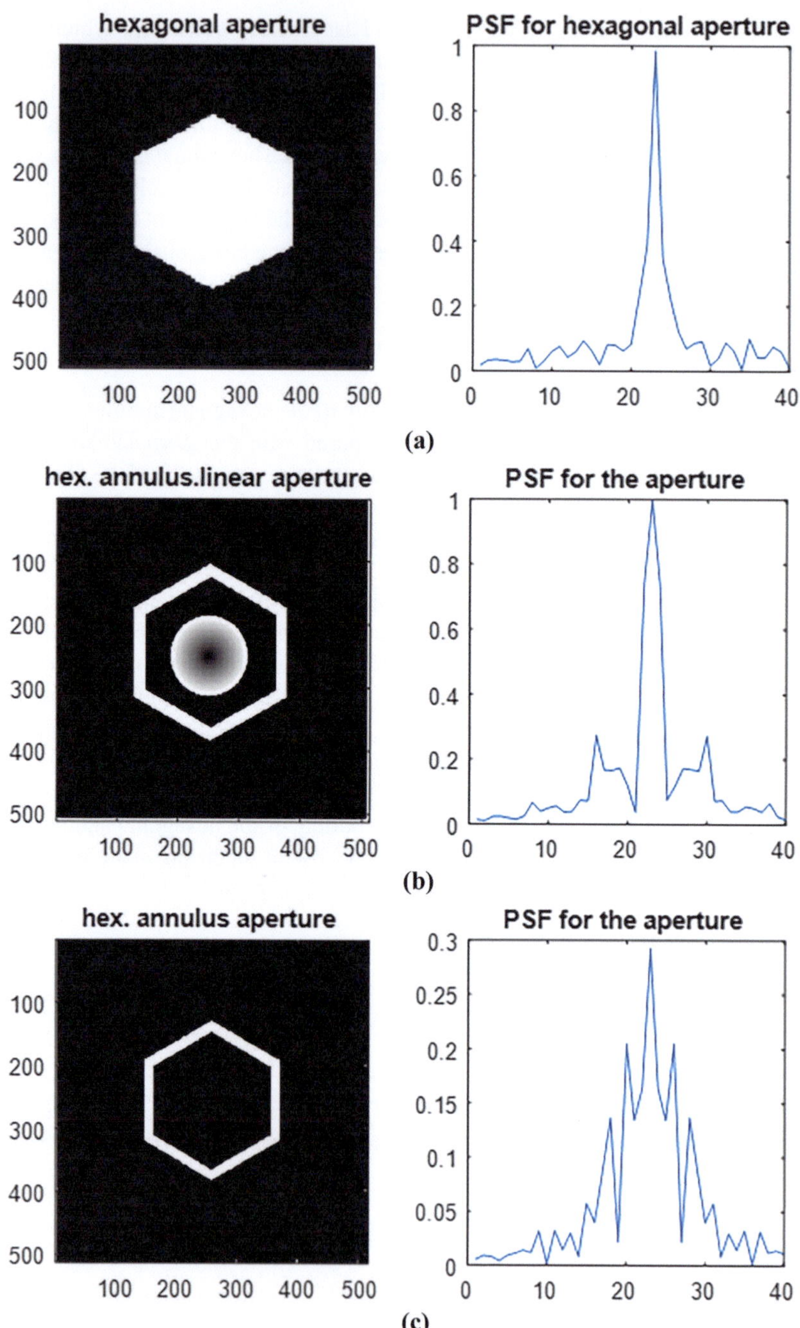

Fig. 5.12 **a** The PSF corresponds to the transparent hexagonal aperture of dimensions 512×512 pixels. The total bandwidth is $= 8$ pixels **b**: The PSF corresponding to the aperture of a hexagonal annulus combined with a central linear distribution of dimensions 512×512 pixels. The total bandwidth is $= 4$ pixels **c** The PSF corresponding to the aperture of the hexagonal annulus of dimensions 512×512 pixels. The total bandwidth is $= 4$ pixels

Table 5.1 Normalized PSF cut-off values corresponding to each modulated aperture

Aperture of radius $= a = b = 1$	PSF cut-off
Transparent hexagonal aperture	0.6347
Hexagonal annulus of width $= 0.2$	0.3808
Hexagonal annulus + linear disc (annular width $= 0.1$)	0.6029
Hexagonal annulus + linear disc (annular width $= 0.2$)	0.6029
Hexagonal annulus + linear disc (annular width $= 0.3$)	0.6029
Hexagonal annulus + linear disc (annular width $= 0.4$)	0.6029
Hexagonal annulus + linear disc (annular width $= 0.5$)	0.6347

Second, the (L/C) aperture has a sharper central peak with a cut-off spatial frequency of 66 pixels compared to the cut-off spatial frequency found at 69 and 70 pixels in the case of (L/B) and (B/W) apertures, respectively. Consequently, optimum resolution is attained using (L/C) aperture.

Referring to Fig. 5.6a, the theoretical curves for the PSF showed moderate resolution for the L/C aperture compared to the resolution corresponding to the linear and conic apertures. The CSLM is applied to the apertures under investigation.

For the second model, we showed that a compromise in resolution and contrast is attained using the hexagonal annulus provided by a central disc of linear distribution. The resolution is equal to the resolution of the hexagonal annulus since the cut-off is equal to 4 pixels. The resolution is invariant for the annular width in the range from 0.1 to 0.4, as shown in Table 5.1. The contrast for the aperture in the second model is better than that obtained for the hexagonal annulus. The optimum contrast for the transparent hexagonal aperture is unity, as expected. Hence, the studied apertures are recommended for imaging using confocal scanning laser microscopy since the detection has a weak intensity, especially for a point detector.

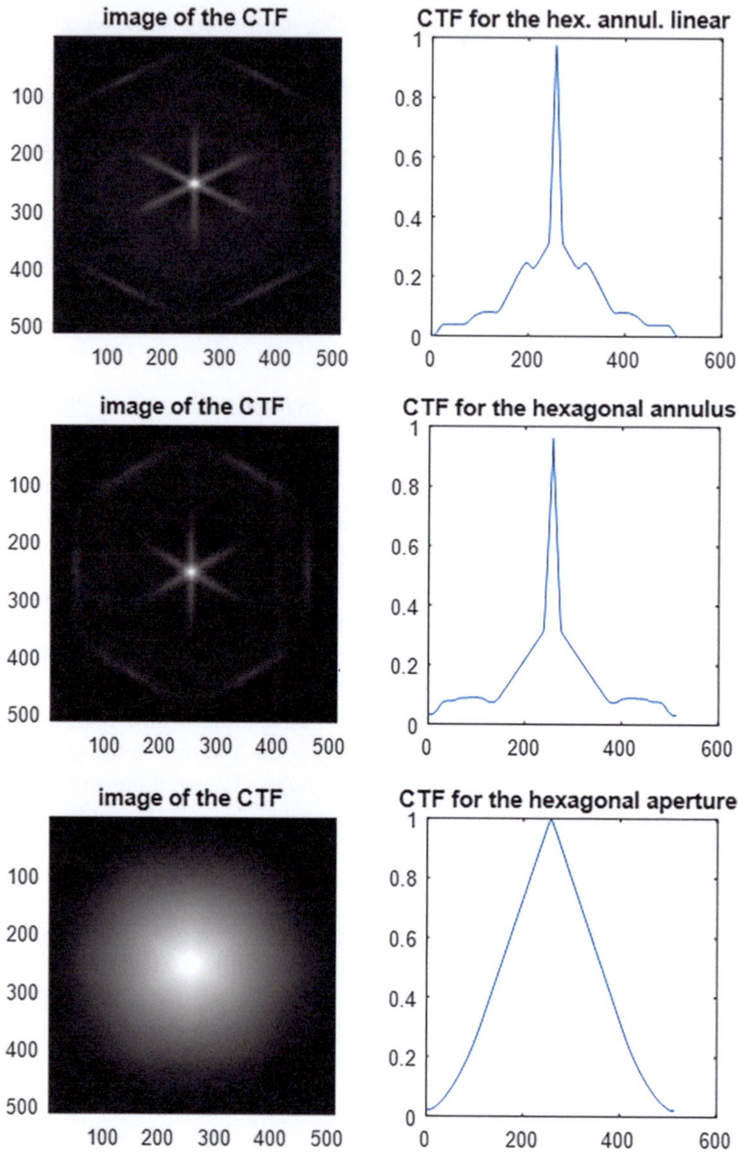

Fig. 5.13 The autocorrelation or the CTF corresponding to the apertures shown in Fig. 5.12a–c

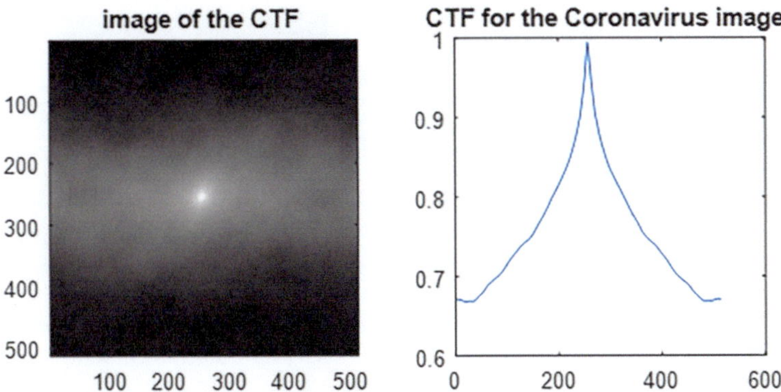

Fig. 5.14 Image of the CTF and its corresponding plot at a height of 256 pixels for the Coronavirus image

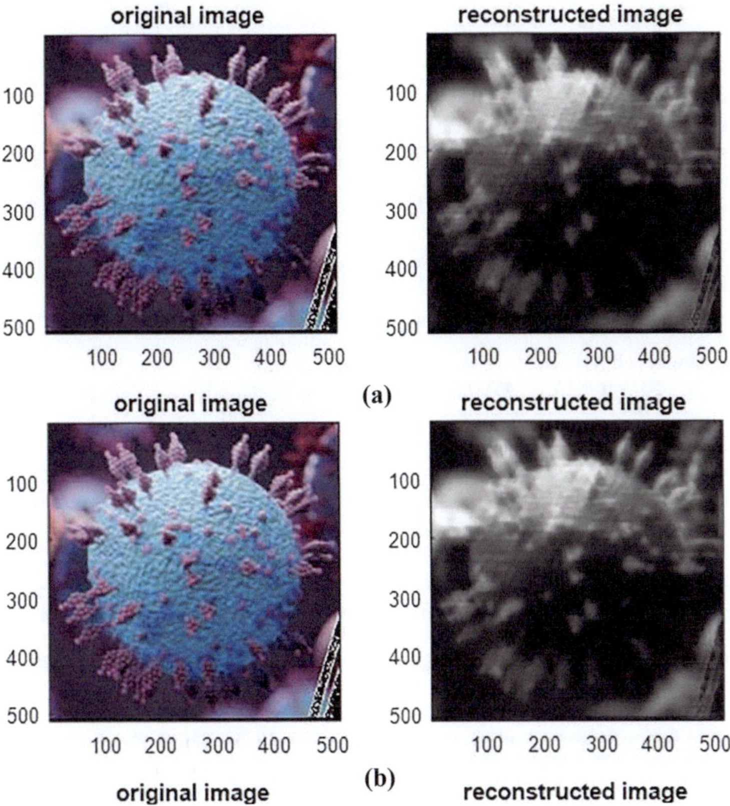

Fig. 5.15 a: The reconstructed image of Coronavirus using the hexagonal annulus combined with a central disc of linear distribution in CSLM **b** The reconstructed image of Coronavirus using the hexagonal annular apertures in CSLM **c** The reconstructed image of Coronavirus using the transparent hexagonal apertures in CSLM

Fig. 5.15 (continued)

References

1. F.A. Hubert, Tschunko, Imaging performance of annular apertures- Line spread functions. Appl. Opt. **17**(7), 1075–1079 (1978)
2. Hubert F. A. Tschunko, Imaging performance of annular apertures. Apodization and modulation transfer functions. Appl. Opt. **18** (22), 3770–3774 (1979)
3. Hubert F. A. Tschunko, Imaging performance of annular apertures. Apodization and point spread functions. Applied Optics **22**, 1, 133–136 (1983). https://doi.org/10.1364/AO.22.000133
4. S.N. Konini and A.V. Ustinov, Sharper focal spot for a radially polarized beam using ring aperture with phase jump. Journal of Engineering (Hindawi Publishing Corporation), ID 512971, 8 (2013). https://doi.org/ https://doi.org/10.1155/2013/512971
5. Hubert F. A Tschunko, Apodization and Image Contrast. Appl. Opt. **18**, 7, 955 (1979)
6. C.J.R. Sheppard, The use of lenses with annular aperture in scanning optical microscopy. Optik **48**(3), 329–334 (1977)
7. C.J.R. Sheppard, T. Wilson, Imaging properties of annular lenses. Appl. Opt. **18**(7), 1058–1063 (1979)
8. A.M. Hamed, J.J. Clair, Image and super-resolution in optical coherent microscopes. Optik **64**, 277–284 (1983)
9. A.M. Hamed, J.J. Clair, Studies on optical properties of confocal scanning optical microscope using pupils with radially transmission distribution. Optik **65**, 209–218 (1983)
10. A.M. Hamed, Resolution and contrast in confocal optical scanning microscope. Opt. and laser technology **16**, 93–96 (1984)
11. A.M. Hamed, Improvement of point spread function (PSF) using linear-quadratic aperture. Optik **131**, 838–849 (2017). https://doi.org/10.1016/j.ijleo.2016.11.201
12. N. Barakat, A.M. Hamed, H. El Chandor, Study of fluid flow using speckle interferometry. Optik **76**, 102–104 (1987)
13. A.M. Hamed, Formation of speckle images formed for diffusers illuminated by modulated apertures (circular obstruction). J. of Modern Opt. **56**, 1633–1642 (2009). https://doi.org/10.1080/09500340903277792
14. A.M. Hamed, Discrimination between speckle images using diffusers modulated by some deformed apertures: Simulations. Opt. Eng. **50**, 1–7 (2011). https://doi.org/10.1117/1.3530085
15. A.M. Hamed, Contrast of laser speckle images using some modulated apertures. Pramana J. Phys. **95**, 122 (2021)
16. A.M. Hamed, Speckle imaging of annular Hermite Gaussian laser beam. Pram. J. Phys. **95** (2021)

17. A.M. Hamed, M.A. Saudi, Computation of surface roughness using optical correlation. Pramana. J. of Phys. **68**, 831–842 (2007)

18. J.J. Clair, A.M. Hamed, Theoretical studies on the optical coherent microscope. Optik **64**, 133–141 (1983)

19. A.M. Hamed, Numerical speckle images formed by diffusers using modulated conical and linear apertures. J. Mod. Opt. **56**(10), 1174–1181 (2009). https://doi.org/10.1080/095003409 02985379

20. C.J.R. Sheppard, The Development of Microscopy for Super-Resolution: Confocal Microscopy, and Image Scanning Microscopy. Appl. Sci. **11**, 8981 (2021). https://doi.org/10.3390/1.119 8981

21. A.M. Hamed, Application of a hexagonal aperture on the confocal scanning laser microscope. Opt. Quant. Electron. **55**, 749 (2023). https://doi.org/10.1007/s11082-023-04920-8

22. A.M. Hamed, Speckle imaging using aperture modulation", Springer Briefs in Applied Sci. and Tech., Chapter 7, 75–88 (2024). https://doi.org/10.1007/978-3-031-58300-1

23. A.M. Hamed, Speckle imaging of annular Hermite Gaussian laser beam. Pramana J. Phys. **95**, 202 (2021). https://doi.org/10.1007/s1.2043-021-02231-9

24. A. Deng, Y. Zheng et al., Improved spatial resolution using focal modulation microscopy with a Tai Chi aperture. Opt. Express **29**(12), 18263–18276 (2021). https://doi.org/10.1364/OE. 426600

25. N. Chen, C.H. Wong, C.J.R. Sheppard, Focal modulation microscopy. Opt. Express **16**(23), 18764–18769 (2008)

26. W. Gong, K. Si, N. Chen, C.J.R. Sheppard, Improved spatial resolution in fluorescence focal modulation microscopy. Opt. Lett. **34**(22), 3508–3510 (2009). https://doi.org/10.1364/OL.34. 003508

27. C. Wu, Y. Zheng et al., Improvements with divided cosine-shaped apertures in confocal microscopy. Opt. Commun. **442**, 71–76 (2019)

Chapter 6
Design of the STRAUBEL Filter in Circular Aperture and Its Application on Confocal Laser Scanning Microscope (CLSM)

6.1 Introduction

In a patent in 1957, Minsky presented the principle of confocal microscopy [1]. The technique scans an object point by point using a focused laser beam to allow for three-dimensional reconstruction. In a conventional microscope, you can only see as far as the light can penetrate, whereas a confocal microscope images one depth level at a time. The Coherent Laser Scanning Microscope (CLSM) works by passing a laser beam through a light source aperture, which is then focused by an objective lens into a small area on the surface of your sample, and an image is built up pixel by pixel by collecting the emitted photons from the fluorophores in the sample. Consequently, in a confocal microscope, the object is illuminated with the focused image of a point source, and the reflected (or transmitted) light intensity is measured with a point detector focused onto the same point of the sample. In practice, the point source is achieved using the laser illumination or an incoherent source with a pinhole. The point detector was achieved using a pinhole in front of the detector. The intensity distribution is the modulus square of the convolution product of the object's complex amplitude and the resultant point spread function (RPSF) corresponding to the CLSM [1–6].

The CLSM is based on a conventional optical microscope, but replaces a lamp with a laser beam focused on the sample. The intensity of the laser light is adjusted by neutral density filters and brought to a set of scanning mirrors that can move them precisely and quickly. One mirror tilts the beam in the X direction, the other in the Y direction. Together, they tilt the beam in a raster fashion or make mechanical scanning. The beam is then brought to the back focal plane of the objective lens, which focuses it onto your sample. If your sample is fluorescent, part of the light will pass back into the objective lens. This light travels backward through the same path that the laser travels. The effect of the scanning mirrors on this light is to produce a spot of light that is not scanning but standing still. This light then passes through a semi-transparent mirror, which takes it away from the laser and toward

© The Author(s), under exclusive license to Springer Nature Switzerland AG 2025
A. M. Hamed, *Image Processing Techniques for Deformed and Sparse Aperture Systems*,
SpringerBriefs in Applied Sciences and Technology,
https://doi.org/10.1007/978-3-032-04921-6_6

the detection system. Any light that emerges from the CLSM's optical system may have an extremely low intensity, and the photomultiplier tube (PMT) is used to detect and amplify this light signal. Photomultipliers can amplify a faint signal around one million times without introducing noise. The output from the PMT is an electrical signal with an amplitude proportional to the initial light output signal. This analog electrical signal is converted to a series of digital numbers by an analog-to-digital (A/D) converter in the computer. As the laser beam moves along the specimen, the detection system constantly samples, converts the PMT output, and displays it on the computer monitor in the correct order. All these complicated steps occur so fast that the display shows a real-time image of the sample. Work seeking to improve the microscope resolution, either conventional or confocal, is made through aperture modulation. For example, linear, quadratic, higher order apertures, graded index apertures, and other combinations of B/W concentric transparent annuli were investigated in [7–14]. Recently, we published a book on modulated apertures and resolution in microscopy [15]. Using slit apertures, rather than pinholes, to construct a confocal imaging system has advantages. The signal level increased, and a line image could be generated in real time if an array of detectors were used. Slit apertures can also be used in a direct-view confocal (tandem scanning) microscope [14]. The factors that influence the resolution of the confocal laser scanning microscope include the diameter of the pinhole in the front of the detector, stray light, and the numerical aperture proposed [16]. In addition, the selection of parameters of the object lens in a point scanning laser confocal microscope was investigated in [17]. Recent work on the resolution in confocal laser microscopes proposed a deep residual neural network algorithm that can effectively improve the image resolution of confocal microscopy in real time [18]. This study contributes to the real-time improvement of the imaging resolution of confocal microscopy and expands the application scenarios of confocal microscopy in biological imaging. In addition, combining adaptive optics with microscope systems can improve the quality of focused beams, reduce aberrations, and effectively increase imaging depth [19–23]. Others used information from adjacent frames to reduce flicker artifacts, thus improving confocal microscopy's resolution and signal-to-noise ratio [24, 25].

In this chapter, a Straubel filter inside the circular aperture was investigated, and we obtained resolution information from the PSF curve. We computed the peak signal-to-noise ratio (PSNR) from the PSF. A relation between the full width at half maximum (FWHM) deduced from the PSF curve and the parameter β is obtained using the Straubel aperture. In addition, the reconstruction of microscopic images is obtained using the Straubel apertures in front of the microscope objectives in the CLSM. The significance of the work is outlined in the novelty of the aperture, where the Straubel aperture is considered.

6.2 Theoretical Analysis

We designed the Straubel filter by considering an aperture used in imaging in the confocal laser scanning microscope (CLSM).

We represent the optical transmittance inside the Straubel aperture as follows:

$$P(\rho) = 1 - \beta\left(\frac{\rho}{\rho_0}\right)^2 ; \left(|(\frac{\rho}{\rho_0})| \leq 1\right) \tag{6.1}$$

ρ is the radial coordinate in the aperture plane (u, v), ρ_0 is the aperture radius, and β is a parameter.

We computed the point spread function by applying the Fourier transform to Eq. (6.1) as follows:

$$h(r) = 2 \int_0^{2\pi} \int_0^{\rho_0} P(\rho,\theta)\, \exp\{-\frac{j2\pi}{\lambda f}(\rho r\cos\theta)\}\rho\, d\rho\, d\theta \tag{6.2}$$

r is the radial coordinate in the Fourier plane (x, y).

The Straubel aperture is independent of the angle θ. It is varied with the radial coordinate ρ.

By substituting Eq. (6.1) with Eq. (6.2), we write the following:

$$h(r) = 4\pi \int_0^{\rho_0} \left[1 - \beta\left(\frac{\rho}{\rho_0}\right)^2\right] \cdot J_0\left(\frac{2\pi}{\lambda f}\rho r\right)\rho\, d\rho \tag{6.3}$$

where J_0 is the Bessel function of zero order.

We rewrite Eq. (6.3) as follows:

$$h(r) = 4\pi(I_1 - I_2) \tag{6.4}$$

$$I_1 = \int_0^{\rho_0} \rho\, J_0\left(\frac{2\pi}{\lambda f}\rho r\right)d\rho \tag{6.5}$$

$$I_2 = \beta \int_0^{\rho_0} \rho\left(\frac{\rho}{\rho_0}\right)^2 J_0\left(\frac{2\pi}{\lambda f}\rho r\right)d\rho \tag{6.6}$$

We replace the reduced coordinate $w = \frac{2\pi}{\lambda f}\rho r$ instead of the radial coordinate r in Eqs. (6.5 and 6.6), we write the following:

$$I_1 = \left(\frac{\lambda f}{2\pi r}\right)^2 \int\limits_0^W w\, J_0(w)dw; \; W = \frac{2\pi r}{\lambda f}\rho_0 \tag{6.7}$$

$$I_2 = \left(\frac{\beta}{\rho_0^2}\right)\left(\frac{\lambda f}{2\pi r}\right)^4 \int\limits_0^W w^3 J_0(w)\, dw \tag{6.8}$$

$$I_1 = \left(\frac{\lambda f}{2\pi r}\right)^2 [W \cdot J_1(W)] = (\rho_0^2).\frac{J_1(W)}{W} \tag{6.9}$$

$$I_2 = \left(\frac{\beta}{\rho_0^2}\right)\left(\frac{\lambda f}{2\pi r}\right)^4 \left[W^3 J_1(W) - 2W^2 J_2(W)\right]; \; \frac{\lambda f}{2\pi r} = \frac{\rho_0}{W}$$

$$I_2 = \beta\rho_0^2\left[\frac{J_1(W)}{W} - 2\frac{J_2(W)}{W^2}\right] \tag{6.10}$$

By substituting Eqs. (6.9 and 6.10) in Eq. (6.4), we obtained the PSF corresponding to the Straubel aperture as follows:

$$h(W) = 4\pi\rho_0^2\left[\frac{J_1(W)}{W}\right] - 4\pi\beta\ \rho_0^2\left[\frac{J_1(W)}{W} - 2\frac{J_2(W)}{W^2}\right] \tag{6.11}$$

The PSF can be rewritten as follows:

$$h(W) = 4\pi\rho_0^2(1 - \beta)\left[\frac{J_1(W)}{W}\right] + 8\pi\beta\ \rho_0^2\left[\frac{J_2(W)}{W^2}\right] \tag{6.12}$$

For $\beta = 0.6$, Eq. (6.12) is reduced to

$$h(W) = \left[\frac{J_1(W)}{W}\right] + 3\left[\frac{J_2(W)}{W^2}\right] \tag{6.13}$$

In Eq. (6.13), we omitted the multiplicative term, which represents the surface area of a circle $= 4\pi\rho_0^2$.

We computed PSNR using the link [26].

To compute the peak signal-to-noise ratio (PSNR), we first calculate the mean squared error (MSE) corresponding to the original input image, $I_1(m, n)$, and the reconstructed image $I_{reconst}(m, n)$ obtained from the CSLM as follows:

$$MSE = \frac{\sum_{M,N}[I_1(m, n) - I_{reconst}(m, n)]^2}{M \times N} \tag{6.14}$$

M and N are the number of rows and columns in the input images. M $=$ N for a square matrix.

The PSNR is computed from the following formula:

$$PSNR = 10\log_{10}\left(\frac{R^2}{MSE}\right) \tag{6.15}$$

In the previous equation, R is the maximum fluctuation in the input image data type.

We calculated the structural similarity (SSIM) index for grayscale image A using ref as the reference image using the following MATLAB code:

$$ssimval = ssim(A, ref)$$

A value closer to 1 indicates better image quality.

6.2.1 Relation Between the Full Width at Half Maximum (FWH) and the Parameter β

Referring to Eq. (6.12), we compute the maximum value at $W = W_0 = 0$. We obtain the following:

$$h_{max}(W_0 = 0) = 4\pi\rho_0^2\left\{0.5(1-\beta) + 2\beta \times 1.25 \times 10^{-4}\right\}$$
$$= 4\pi\rho_0^2\{0.5 - 0.4999\} \tag{6.16}$$

$$h_{max}(W = 0) \approx 2\pi\rho_0^2(1-\beta) \tag{6.17}$$

To obtain the FWHM or (ΔW), we take the ratio $\frac{h(W)}{h_{max}} = \frac{1}{2}$. Hence, we obtain the following relation for the FWHM as a function of the parameter β:

$$(1-\beta)\Delta W^2 - 8(1-\beta)\Delta W J_1\left(\frac{\Delta W}{2}\right) - 32\beta\, J_2\left(\frac{\Delta W}{2}\right) = 0 \tag{6.18}$$

Some results of the FWHM at different parameters β are reported in the following Table 6.1.

Table 6.1 Relation between the FWHM and the parameter β

FWHM	The parameter β
4	0
5	0.2638
6	0.5589
7	0.7379
8	0.8541
9	0.9334
10	0.9883

6.3 Results and Discussion

We fabricated a computerized Straubel aperture of radius $= 64$ pixels in a matrix of dimensions 1024×1024 pixels, where the parameter $\beta = 0.6$. The optical transmittance inside the Straubel aperture is plotted as in Fig. 6.1a. A Straubel aperture of radius $= 128$ pixels and its optical transmittance is shown in Fig. 6.1b.

We computed the normalized PSF corresponding to the Straubel aperture of different parameters β using the fast Fourier transform (FFT). We showed the PSF plots in Fig. 6.2a–d. The FWHM computed from the PSF graph is 10, 11, 12, and 13 pixels, respectively. The aperture radius $= 64$ pixels for all the curves. Referring to Fig. 6.2a–d, we showed that the FWHM increases with the increase in the parameter β. Consequently, the FWHM decreased, and hence the resolution improved for low β. The Straubel aperture for $\beta = 1.0$ is equivalent to the inverted quadratic aperture.

For comparison, the normalized PSF corresponds to the circular aperture of radius $= 64$ pixels shown in Fig. 6.2e. The FWHM computed from the PSF graph is equal to 10 pixels, where the peak is at 66 pixels. We showed equal values of FWHM in the case of circular aperture and Straubel aperture at $\beta = 0.4$, Fig. 6.2a.

Another comparison of the normalized PSF corresponding to the quadratic aperture is shown in Fig. 6.2f. The aperture radius is equal to 64 pixels. The FWHM computed from the PSF graph is 8 pixels, where the peak is 66 pixels. In this case, we showed that the cut-off is the lowest, giving a sharp FWHM $= 8$ pixels compared to the Straubel and circular apertures. Hence, the resolution in the case of quadratic apertures is the optimum, while the Straubel and circular apertures have the same resolution. In addition, the contrast using the Straubel aperture is changed by the parameter β. The PSF in Fig. 6.2 is computed using the Fast Fourier transform (FFT).

Referring to the results of the FWHM at different values (β) obtained from Eq. (6.18), we showed that the FWHM in the case of the Straubel aperture increases

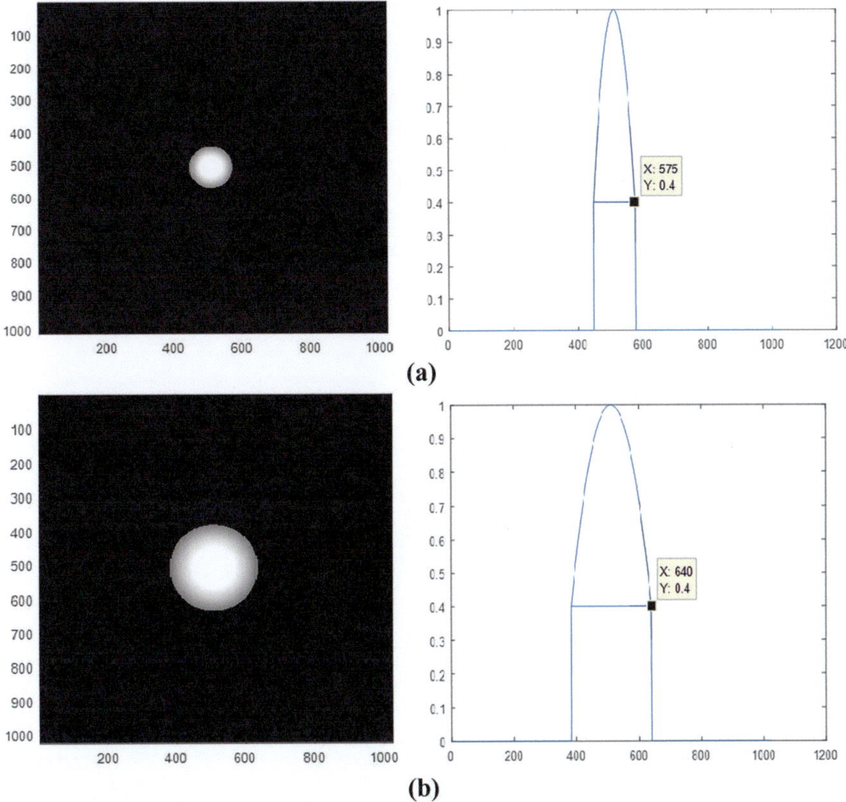

Fig. 6.1 **a** Straubel aperture of radius $= 64$ pixels in a matrix of dimensions 1024×1024 pixels and the corresponding transmittance distribution curve inside the aperture. The parameter $\beta = 0.6$. **b** Straubel aperture of radius $= 128$ pixels in a matrix of dimensions 1024×1024 pixels and the corresponding transmittance distribution curve inside the aperture. The parameter $\beta = 0.6$

gradually with the parameter β. In addition, the FWHM for the Straubel aperture is smaller than that of a circular aperture at $\beta \leq 0.9334$.

We computed the normalized PSF using the derived Eq. (6.13), for the Straubel aperture for the parameter $\beta = 0.6$. We plotted the PSF in Fig. 6.3a. Two comparative PSF curves are plotted in Fig. 6.3b and c, corresponding to the circular and quadratic apertures. The aperture radius of 2 μm in all PSF plots is shown in Fig. 6.3a–c. Referring to Fig. 6.3, the Straubel aperture has FWHM $= 109$ μm compared to the circular and quadratic apertures of FWHM $= 101$ and 93 μm, respectively. Consequently,

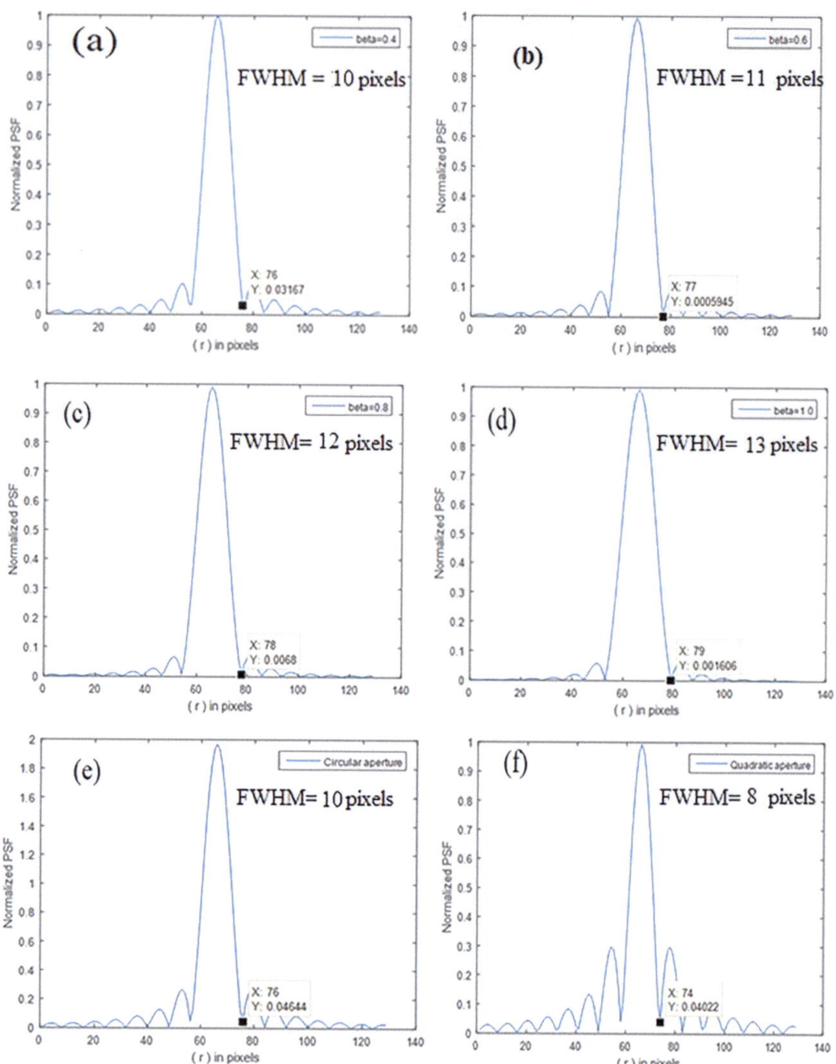

Fig. 6.2 Normalized PSF corresponding to the Straubel aperture of radius = 64 pixels, where the parameter β = 0.4, 0.6, 0.8, and 1.0 compared with the PSF for circular and quadratic apertures

the quadratic aperture has a sharp FWHM compared to the Straubel, and the circular apertures agree with the FWHM results shown using the FFT.

The input image (cancer malignant cells) used in the reconstruction is shown in Fig. 6.4.

We obtained the reconstructed image corresponding to the original cancer malignant cells using the CSLM provided with Straubel apertures for the two objectives

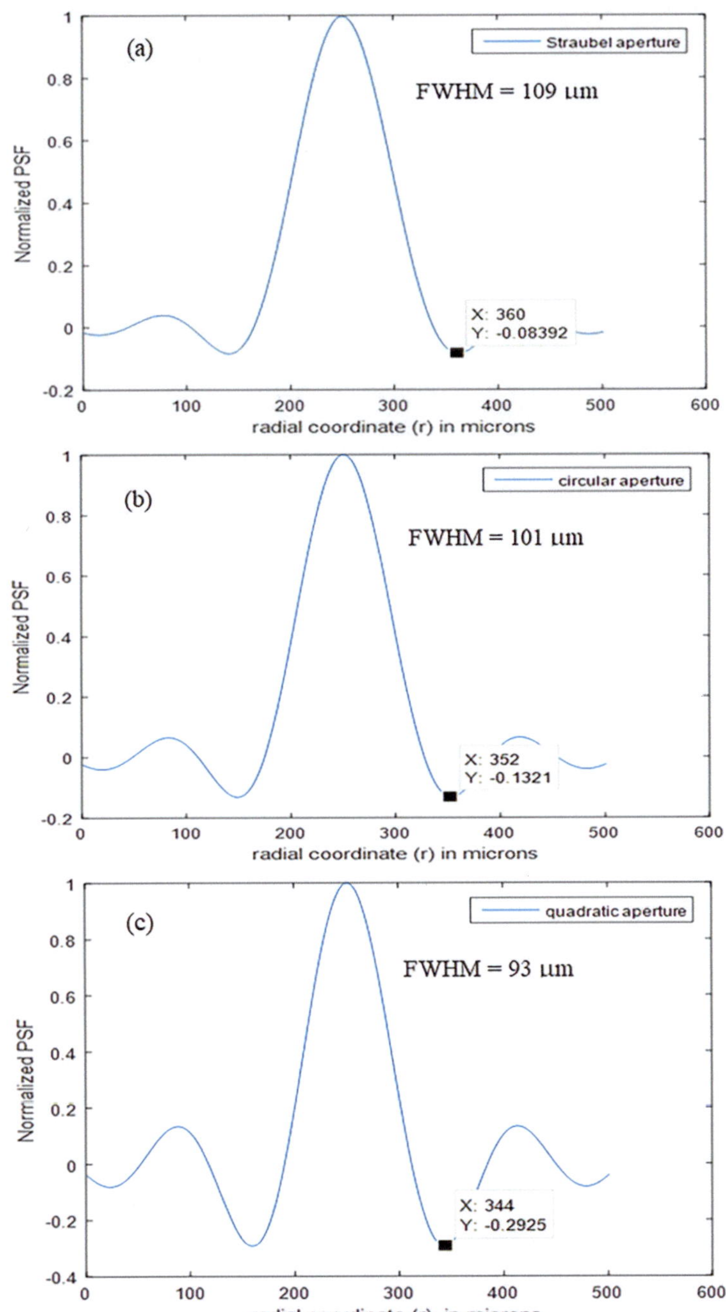

Fig. 6.3 Normalized PSF corresponding to different apertures of radius $= 2$ μm. The FWHM computed from the PSF graph is equal to 109 μm in (**a**), 101 μm in (**b**), and 93 μm in (**c**), where the peak is at 251 μm

Fig. 6.4 The Input image (cancer malignant cells) is used in the reconstruction. Reference the link [27]

of the microscope, as in Fig. 6.5a. The contrast of the reconstructed image $= 0.6415$, and the peak signal-to-noise ratio (PSNR) $= 23.23$.

The reconstructed image obtained using one circular aperture and another Straubel is shown in Fig. 6.5b, and the PSNR $= 21.58$. The contrast decreased to 0.6381. For two similar circular apertures, the reconstructed image shown in Fig. 6.5c, and the contrast remains $= 0.6381$. For two apertures of quadratic distribution, the reconstructed image shown in Fig. 6.5d, the corresponding contrast decreased to 0.5999, and the peak signal-to-noise ratio (PSNR) $= 22.48$. The highest value of PSNR $= 23.23$ corresponds to two symmetric Straubel apertures, and the lowest value of PSNR $= 21.58$ for two different apertures of circular and Straubel apertures. Similar values of PSNR are computed in the case of two symmetric circular apertures or two symmetric quadratic apertures. Hence, the PSNR for the reconstructed images from the CSLM is affected by the amplitude distribution of the apertures.

Referring to the results in the table, the structural similarity (SSIM) index for the reconstructed grayscale image increased for the two symmetric circular apertures decreased for the other combinations of modulated apertures.

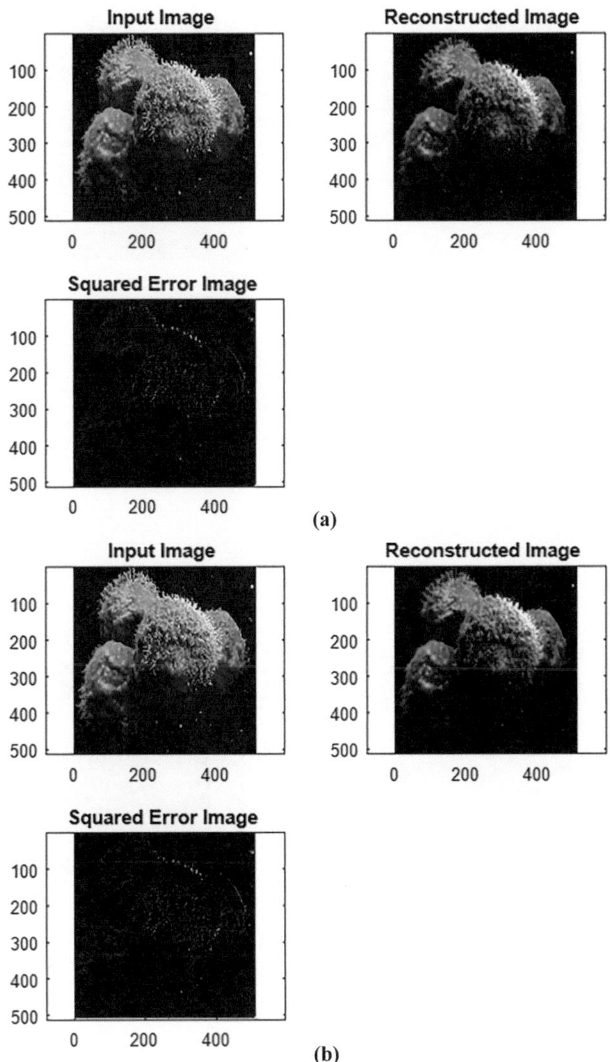

Fig. 6.5 Reconstructed image corresponding to the original cancer cells using the CSLM provided with Straubel apertures for the two objectives of the microscope. The contrast of the reconstructed image $= 0.6415$. The peak signal-to-noise ratio (PSNR) $= 23.23$. **b** Reconstructed image using the CSLM provided with Straubel apertures for the first objective of the microscope and the second with a circular aperture. The contrast of the reconstructed image $= 0.6381$. The PSNR is 21.58. **c** Reconstructed image using the CSLM provided with both circular apertures for the two objectives of the microscope. The contrast of the reconstructed image $= 0.6381$. The PSNR is 22.49. **d** Reconstructed image using the CSLM provided with both apertures of quadratic distributions for the two objectives of the microscope. The contrast of the reconstructed image $= 0.5999$. The PSNR is 22.48

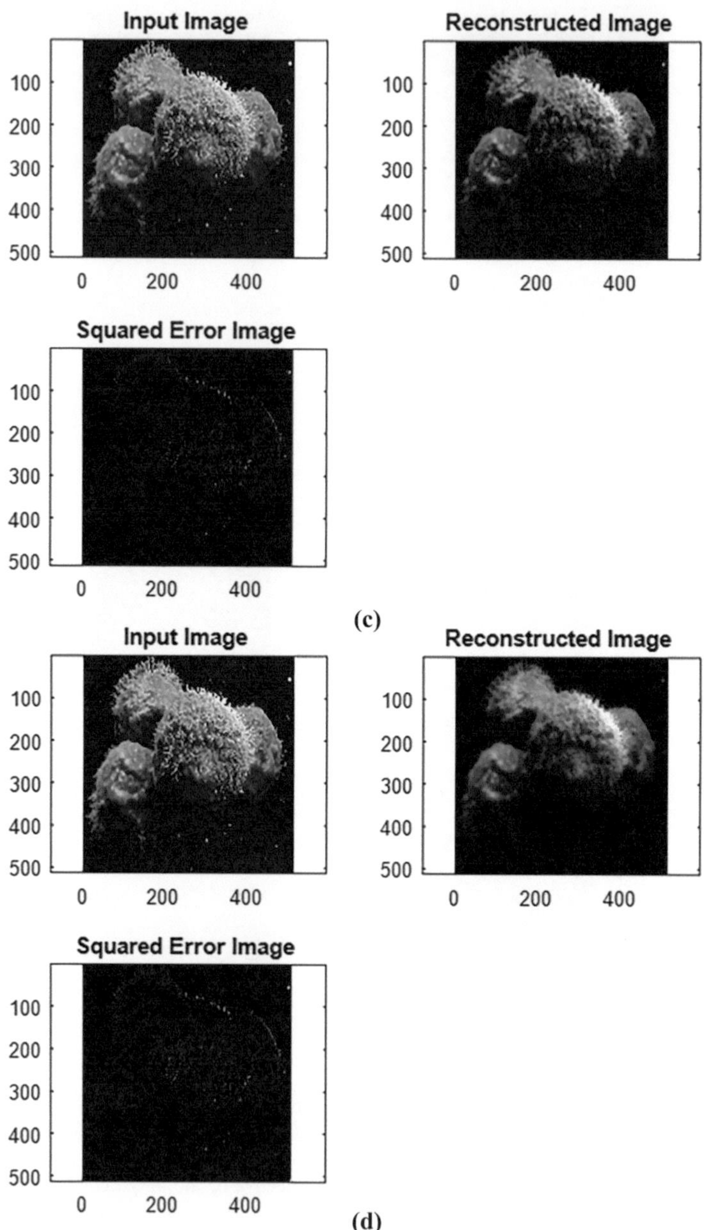

Fig. 6.5 (continued)

Table 6.2 The contrast, SSIM, and PSNR corresponding to the reconstructed images are plotted in the table

The two apertures of the CSLM	Contrast	SSIM	PSNR
1-Two straubel apertures	0.6415	0.1557	23.23
2-One straubel and the other circular aperture	0.6381	0.1532	21.58
3-Two circular apertures	0.6381	0.2105	22.49
4-Two quadratic apertures	0.5999	0.1817	22.48

6.4 Conclusion

We deduced that the Straubel aperture has a resolution for the parameter $\beta = 0.4$, equal to the resolution in the case of a circular aperture. In addition, the PSF shape depends on the parameter β, and the FWHM increases with the increase of (β). We considered that the Straubel aperture is equal to the inverted quadratic aperture for the parameter (β) equal to unity.

The reconstructed images using the confocal microscope provided with both Straubel apertures showed better contrast and improved PSNR compared to the reconstructed images corresponding to the combination of circular and Straubel apertures. We deduced that the resolution improvement was obtained using quadratic apertures, shown by the PSF plots. It leads to a decrease in the image contrast and PSNR. Consequently, the Straubel apertures are useful in improving image contrast in confocal microscopy.

References

1. M. Minsky, Microscopy apparatus. United States Patent Office. Filed Nov. 7, 1957, granted Dec. 19, 1961. Patent No. 3, 013, 467 (1961)
2. C.J.R. Sheppard, A. Choudhury, Image formation in the scanning microscope. Optica Acta: J. Modern Optics **24**, 1051–1073 (1977)
3. C.J.R. Sheppard, Scanned imagery. J. Phys. D, **19**, 2077–2084 (1986). https://doi.org/10.1088/00223727/19/11/007
4. C.J.R. Sheppard, T. Wilson, Image formation in scanning microscopes with partially coherent source and detector. Optica Acta: J. Modern Optics **25**, 315–325 (1978)
5. C.J.R. Sheppard, H.J. Matthews, Imaging in high-aperture optical systems. J. Opt. Soc. Am. A **4**, 1354–1360 (1987)
6. I.J. Cox, C.J.R. Sheppard, Information capacity and resolution in an optical system. J. Opt. Soc. Am. **3**, 1152–1158 (1986)
7. J.J. Clair, A.M. Hamed, Theoretical studies on optical coherent microscopes. Optik **64**(2), 133–141 (1983)
8. A.M. Hamed, J.J. Clair, Image and super-resolution in optical coherent microscopes. Optik **64**, 277–284 (1983)

9. A.M. Hamed, J.J. Clair, Studies on optical properties of confocal scanning optical microscope using pupils with radially transmission distribution. Optik **65**, 209–218 (1983)

10. A.M. Hamed, Resolution and contrast in confocal optical scanning microscope. Opt. Laser Technol. **16**, 93–96 (1984)

11. A.M. Hamed, Study of the graded index and truncated apertures using speckle images. Precis. Instrum. Mech. (PIM) **3**, 144–152 (2014)

12. A.M. Hamed, T. Al-Saeed, Image analysis of modified Hamming aperture: application on confocal microscopy and holography. J. Modern Opt. **62**, 801–810 (2015)

13. A.M. Hamed, Improvement of point spread function (PSF) using linear-quadratic aperture. Optik **131**, 838–849 (2017)

14. C.J.R. Sheppard, X.Q. Ma, Confocal microscopes with slit apertures. J. Modern Optics **35**, 1169–1185 (1988)

15. A.M. Hamed, Modulated apertures and resolution in microscopy. Springer Briefs in Applied Sciences and Technology, Springer Nature (2023). https://link.springer.com/book/10.1007/978-3-031-47552-8

16. M. Chang, P. Zhang, J. Sun, X. Zhang, Factors influencing the resolution of the confocal laser scanning optical microscope. Adv. Optoelectron. Micro/Nano-Opt., (2010). https://doi.org/10.1109/AOM.2010.5713585

17. L. Bing, W. Zhao, L. Ming-Zhou, Selection for parameters of object lens in point scanning laser confocal microscope. Tool Engineering **38**, 51–53 (2004)

18. Z. Cui, Y. Xing et al., Real-time resolution enhancement of confocal laser scanning microscopy via deep learning.Photonics **11**, 10 (2024). https://www.mdpi.com/2304-6732/11/10/983

19. C. Stockbridge, Y. Lu et al., Focusing through dynamic scattering media. Opt. Express **20**, 15086–15092 (2012)

20. I. Galaktionov, A. Nikitin et al., Focusing of a laser beam passed through a moderately scattering medium using a phase-only spatial light modulator. Photonics **9**, 296 (2022)

21. Q. Katz, E. Small, Y. Guan, Y. Silberberg, Noninvasive nonlinear focusing and imaging through strongly scattering turbid layers. Optica **1**, 170–174 (2014)

22. T.R. Hillman, T. Yamauchi et al., Digital optical phase conjugation for delivering two-dimensional images through turbid media. Sci. Rep. **3**, 1909 (2013)

23. X. Tao, B. Fernandez et al., Adaptive optics confocal microscopy using direct wavefront sensing. Opt. Lett. **36**, 1062–1064 (2011)

24. L. Fang, F. Monroe et al., Deep learning-based point-scanning super-resolution imaging. Nat. Methods **18**, 406–416 (2021)

25. B. Huang, J. Li et al., Enhancing image resolution of confocal fluorescence microscopy with deep learning. Photonix **4**, 2 (2023)

26. PSNR—Compute peak signal-to-noise ratio (PSNR) between images—Simulink

27. https://www.gettyimages.com/detail/photo/cancer-malignant-cells-royalty-free-image/1372020529

Chapter 7
Investigation of Modulated Cardiac Apertures and Their Applications in Confocal Laser Scanning Microscope

7.1 Introduction

The subject of a confocal laser scanning microscope was suggested in the patent of Minsky [1]. The optical system of this confocal arrangement, working in transmission, comprises two microscope objectives arranged in tandem. A mechanically scanned object is placed in the common short focus of the objectives. A point detector is placed in the imaging plane to detect the image. The image was built on the oscilloscope screen during the synchronization of the electronic scanning with the mechanical scanning of the object. The patent was followed by numerous theoretical studies on the confocal microscope, Sheppard et al. [2–10]. Others showed the effect of annular width on the focal depth [11, 12].

Recently, image processing of medical images based on speckle imaging technique and its contrast have been presented in [14–20], while the image processing using the confocal scanning laser microscope provided with modulated apertures is outlined in [21–24]. Another recent work using the confocal microscope was presented in [25–27]. Compressive light-field microscopy for three-dimensional neural activity recording is shown in [25]. A remodeling of cardiac tissue was analyzed, and a comprehensive approach based on confocal microscopy, leading to three-dimensional reconstruction, was presented in [26]. Catheterized Fiber-Optics Confocal Microscopy of the Beating Heart In Situ was investigated in [27].

We believe that the cardiac aperture is a new modulating aperture. Hence, we present this study to investigate these apertures. We considered four different models of cardiac apertures. We computed the point spread function (PSF) corresponding to each model. We compared the PSF results with those corresponding to the transparent cardiac aperture. In addition, we studied the effect of annular width on the PSF. Finally, we apply these modulated cardiac apertures on the confocal microscope.

A. M. Hamed, *Image Processing Techniques for Deformed and Sparse Aperture Systems*,
SpringerBriefs in Applied Sciences and Technology,
https://doi.org/10.1007/978-3-032-04921-6_7

7.2 Analysis

In the present analysis, we consider the B/W concentric aperture of cardioid shape.

The transparent cardiac aperture in Cartesian coordinates is represented as follows [19, 28]:

$$p(x, y) = 1; \ \frac{(x^2 + y^2 - 2ax)^2}{4a^2(x^2 + y^2)} \leq 1, and \ p(x, y) = 0; \ \text{otherwise.} \tag{7.1}$$

In polar coordinates, it is represented as follows:

$$p(\rho, \theta) = 1; \ \rho(\theta) = 2a[1 - \cos(\theta)] \tag{7.2}$$

a is a constant, while the radius ρ changed with the angle θ.

The constant a is deduced from Eq. (7.1) as follows:

$$a = \frac{x^2 + y^2}{2[x + \sqrt{x^2 + y^2}]} \tag{7.3}$$

Equation (7.3) only holds for the variables (x) and (y) that lie on the contour of the cardioid; it does not hold for any other (x), (y) as suggested in Eq. 7.3.

The parametric equations of a planar cardioid are written as follows:

$$x = 2a \cos(\theta)[1 - \cos(\theta)], \ and \ y = 2a \sin(\theta)[1 - \cos(\theta)] \tag{7.4}$$

Now, the cascaded six B/W concentric cardiac annuli, considering the outer annulus is transparent and lasts with a black annulus near the center, are represented as follows:

$$\Delta p_1(\rho, \theta) = 1; \ \Delta \rho_1(\theta) = \rho_1(\theta) - \rho_2(\theta) = 2(a_1 - a_2)[1 - \cos(\theta)] \tag{7.5}$$

$$\Delta p_2(\rho, \theta) = 1; \ \Delta \rho_2(\theta) = \rho_3(\theta) - \rho_4(\theta) = 2(a_3 - a_4)[1 - \cos(\theta)] \tag{7.6}$$

$$\Delta p_3(\rho, \theta) = 1; \ \Delta \rho_3(\theta) = \rho_5(\theta) - \rho_6(\theta) = 2(a_5 - a_6)[1 - \cos(\theta)] \tag{7.7}$$

where $a_1 = a$ in Eq. (7.3):

$$a_1 = \frac{x^2 + y^2}{2[x + \sqrt{x^2 + y^2}]}, a_2 = \frac{(x - \Delta x)^2 + y^2}{2[(x - \Delta x) + \sqrt{(x - \Delta x)^2 + y^2}]},$$

$$a_3 = \frac{(x - 2\Delta x)^2 + y^2}{2[(x - 2\Delta x) + \sqrt{(x - 2\Delta x)^2 + y^2}]} \quad ,$$

$$a_4 = \frac{(x - 3\Delta x)^2 + y^2}{2[(x - 3\Delta x) + \sqrt{(x - 3\Delta x)^2 + y^2}]}$$

$$a_5 = \frac{(x - 4x)^2 + y^2}{2[(x - 4x) + \sqrt{(x - 4x)^2 + y^2}]},$$

$$a_6 = \frac{(x - 5x)^2 + y^2}{2[(x - 5x) + \sqrt{(x - 5x)^2 + y^2}]} \tag{7.8}$$

Hence, the B/W cardiac aperture of a finite number of zones N is represented as follows:

$$p(\rho, \theta) = \Delta p_1(\rho, \theta) + \Delta p_2(\rho, \theta) + \Delta p_3(\rho, \theta) + \cdots + \Delta p_N(\rho, \theta) \tag{7.9}$$

where $\Delta p_1 = P_1 - P_2$, $\Delta p_2 = P_3 - P_4$, $\Delta p_3 = P_5 - P_6$, etc. $\tag{7.10}$

The PSF for the B/W cardiac aperture is computed by applying the Fourier transform to Eq. (7.9) as follows:

$$h(r) = \int_0^{2\pi} \int_0^{\rho} P(\rho, \theta) \exp\{-\frac{j2\pi}{\lambda f}(\rho r \cos\theta)\}\rho \, d\rho \, d\theta \tag{7.11}$$

where ρ is a variable dependent on the angle (θ), represented by Eq. (2.2), and r is the radial coordinate in the Fourier plane, where $r = \sqrt{u^2 + v^2}$.

In the next section, we computed the PSF corresponding to the B/W cardiac apertures Eq. (7.11) using the fast Fourier transform (FFT). We computed the transparent cardiac aperture from Eqs. (7.1) and (7.2), and we computed the B/W cardiac aperture using Eqs. (7.5–7.10). The radius ρ is not a constant like in a circular aperture; it varies according to Eq. (7.2), in the range [0, 2a].

The algorithm used to reconstruct the input image is outlined in [29].

7.3 Results

We constructed a cardiac aperture composed of eight zones of black-and-white (B/W) annuli of equal width = 32 pixels and plotted it in Fig. 7.1a. The maximum radii are equal to 32, 64, 96, 128, 160, 192, 224, and 256 pixels in a matrix of dimensions 1024 × 1024 pixels. The PSF corresponding to the B/W cardiac aperture, shown in Fig. 7.1a, versus x-coordinate in pixels at constant y = 512 pixels is plotted using the FFT as in Fig. 7.1b. From the graph, the total BW is 6 pixels, and the full width at half maximum (FWHM) is 2 pixels. The plots for the other models are shown in Figs. 7.2, 7.3, and 7.4. The cardiac aperture of linear distribution and its PSF plots are shown in Fig. 7.2a and b. The total BW and FWHM are the same as for the first model and equal to 6 and 2 pixels, respectively. In the third model, a modulated cardiac aperture composed of a linear distribution surrounded by an annulus and its corresponding PSF plots are shown in Fig. 7.3a and b. The fourth model of the annular cardiac aperture of width = 32 pixels along the x-axis and the corresponding PSF plots are shown in Fig. 7.4a and b. The comparative transparent cardiac aperture and its PSF plots are shown in Fig. 7.5a and b.

For all plots shown in Figs. 7.1, 7.2, 7.3, 7.4 and 7.5, the maximum radius is 256 pixels in a matrix of dimensions 1024 × 1024 pixels. We used MATLAB code to construct all the apertures and the corresponding PSF plots.

The normalized coherent transfer function (CTF) corresponding to the B/W cardiac aperture is computed using FFT operations and is shown in Fig. 7.6a. The CTF image is shown on the L.H.S., and the corresponding line plot at constant y = 512 pixels is shown on the R.H.S. Similar plots for the CTF corresponding to the other modulated apertures are shown in Fig. 7.6b, c, d, and e.

The reconstructed images corresponding to the original image using the CSLM are obtained [29]. We plotted them in Fig. 7.7a–e. In Fig. 7.7a, we considered eight B/W concentric cardiac annuli for both microscope objectives. In Fig. 7.7b, the apertures have a linear distribution. In Fig. 7.7c, an annulus surrounding the linear distribution is considered. In Fig. 7.7d, an annular aperture is considered. In addition, transparent cardiac apertures are used for comparison in Fig. 7.7e.

7.4 Discussion

Referring to the PSF plots, shown in Figs. 7.1, 7.2, 7.3, 7.4 and 7.5, the central peak has a similar total bandwidth = 6 pixels; except for the annular cardiac aperture and the aperture of linear distribution, the PSF has a narrower bandwidth = 4 pixels. In addition, the secondary peaks of the PSF plots for all the models are stringent compared with those of a transparent cardiac aperture. Hence, the suggested models are considered useful for imaging extended objects. We showed that the PSF profile depends on the aperture shape as expected.

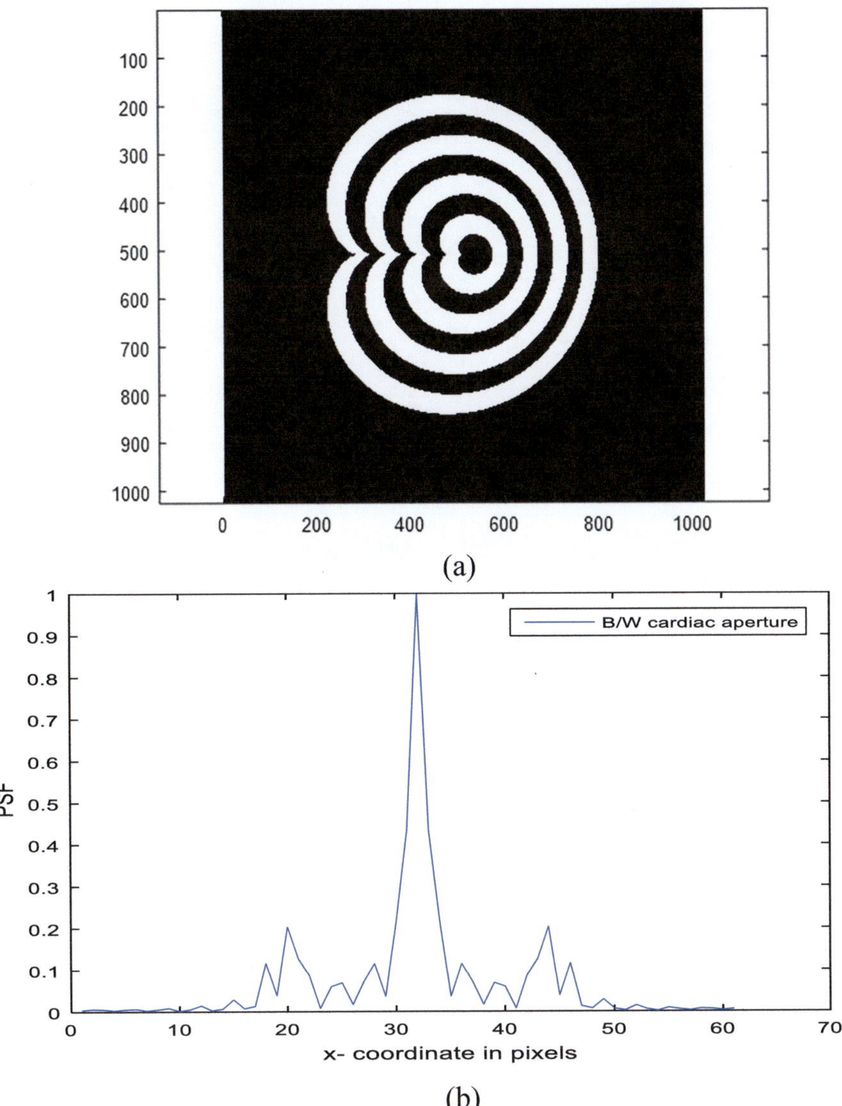

(a)

(b)

Fig. 7.1 **a** A cardiac aperture composed of eight zones of black-and-white (B/W) annuli of equal width $= 32$ pixels. The maximum radii are equal to 32, 64, 96, 128, 160, 192, 224, and 256 pixels in a matrix of dimensions 1024×1024 pixels **b** The PSF corresponding to the B/W cardiac aperture, shown in Fig. 7.1a, versus x-coordinate in pixels at constant $y = 512$ pixels. The total BW $= 6$ pixels and the FWHM $= 2$ pixels

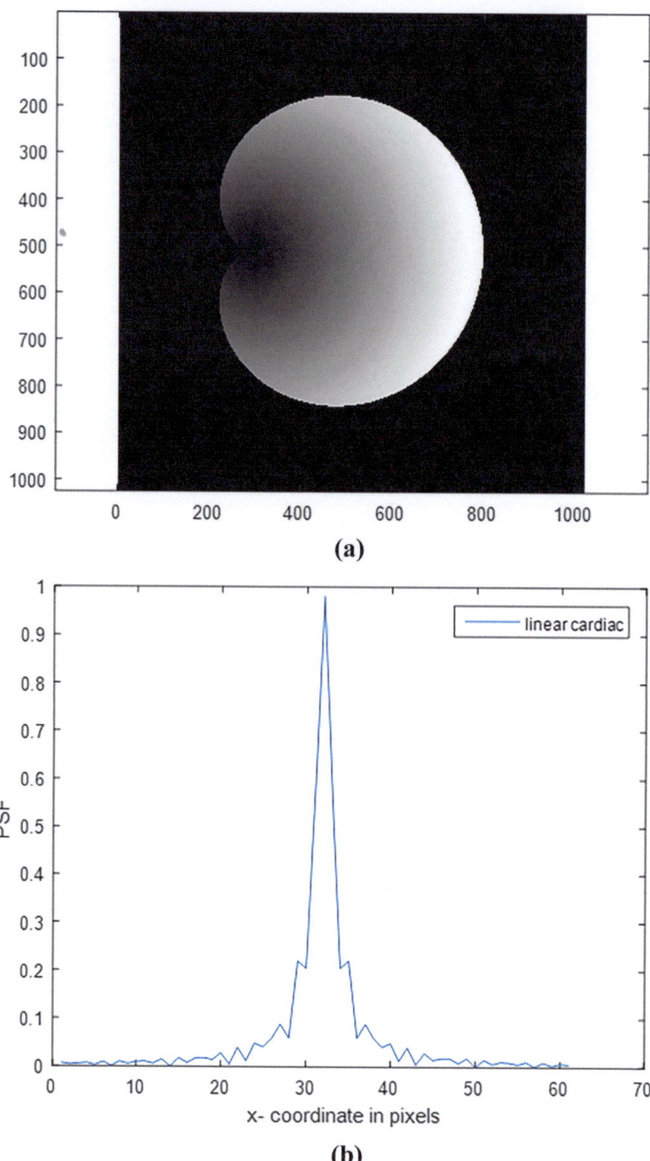

(a)

(b)

Fig. 7.2 **a** A cardiac aperture of a linear distribution. The maximum radius is 256 pixels in a matrix of dimensions 1024×1024 pixels **b** The PSF corresponding to the cardiac aperture of linear distribution, shown in Fig. 7.2a, versus x-coordinate in pixels at constant $y = 512$ pixels. The total $BW = 4$ pixels, and the $FWHM = 2$ pixels

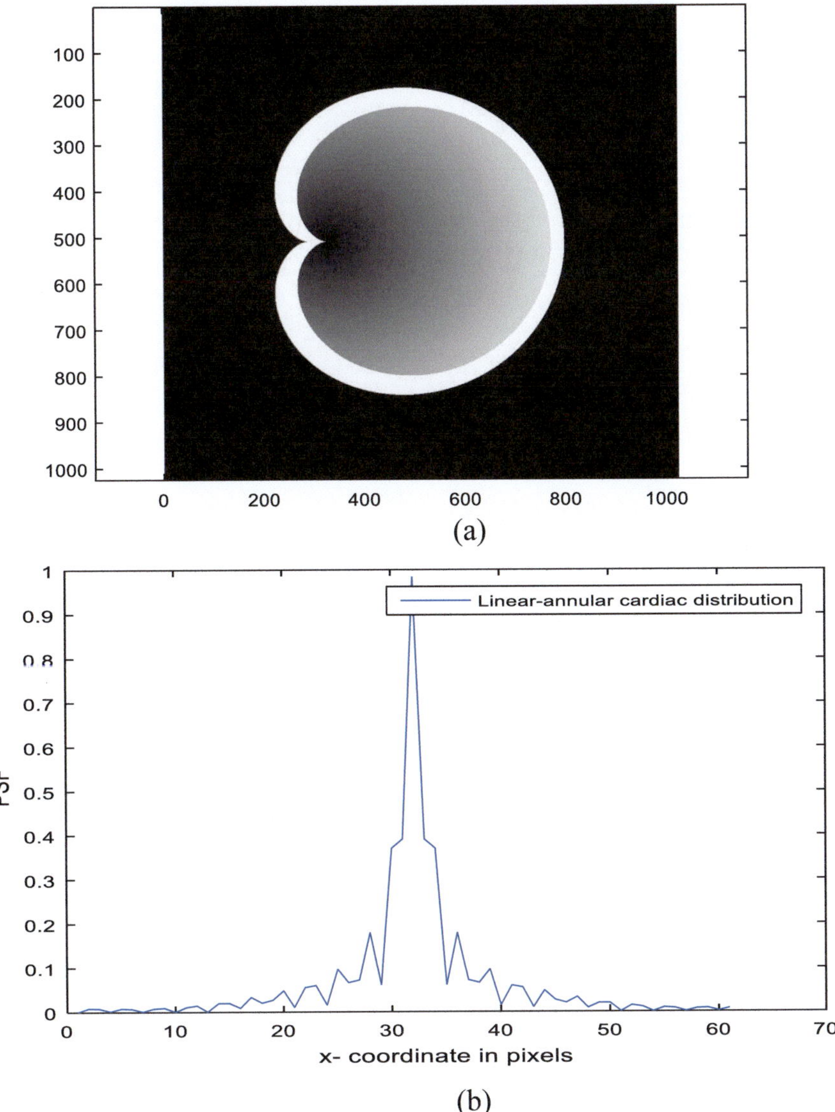

(a)

(b)

Fig. 7.3 **a** A modulated cardiac aperture of a linear distribution surrounded by an annulus. The maximum radius of the whole aperture is 256 pixels, and the annular width = 32 pixels in a matrix of dimensions of 1024 × 1024 pixels **b** The PSF corresponding to the linear-annular cardiac aperture, shown in Fig. 7.3a, versus x-coordinate in pixels at constant y = 512 pixels. The width of the annulus = 32 pixels. The total BW = 6 pixels, and the FWHM = 2 pixels

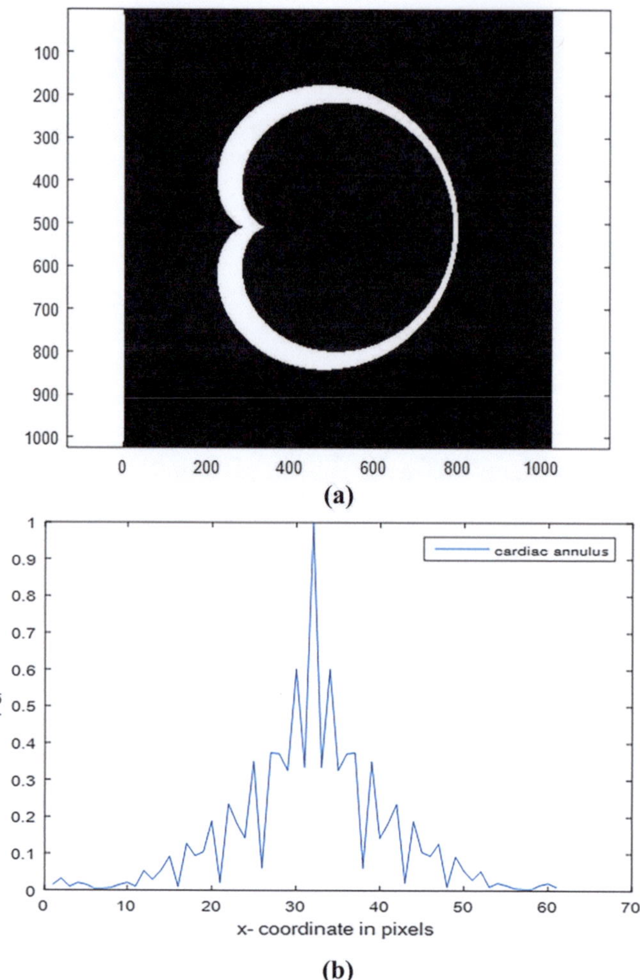

Fig. 7.4 **a** A cardiac annulus of width = 32 pixels along the x-axis. The maximum radius is 256 pixels in a matrix of dimensions 1024×1024 pixels **b** The PSF corresponding to the annular cardiac aperture versus x-coordinate in pixels at constant y = 512 pixels. The total BW = 2 pixels, and the FWHM = 1 pixel

The CTF in the case of B/W cardiac apertures has a distinguished shape characterized by fringing around the center due to the aperture structure. In addition, the annular aperture has a sharp peak as expected from its geometry. The other models with linear, linear-annular, and transparent cardiac apertures have simple decaying shapes.

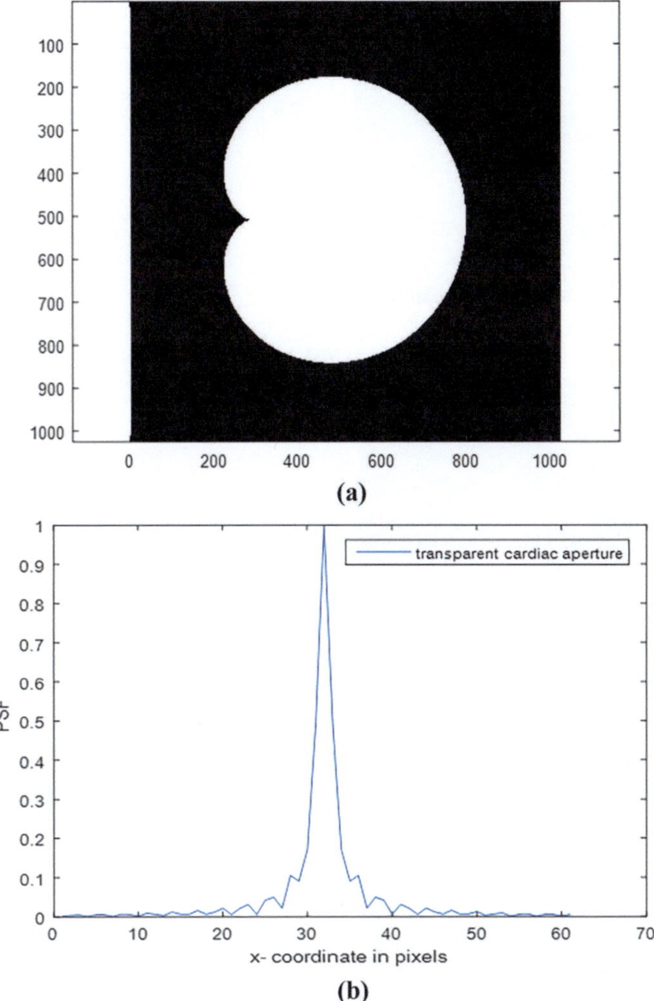

Fig. 7.5   a A transparent cardiac aperture of maximum radius $= 256$ pixels in a matrix of dimensions 1024×1024 pixels **b** The PSF corresponding to the transparent cardiac aperture versus x-coordinate in pixels at constant $y = 512$ pixels. The total BW $= 6$ pixels, and the FWHM $= 2$ pixels

The reconstructed images from the CSLM have different contrasts depending on the transmitted intensity from the apertures, considering the same object. In addition, the resolution of the reconstructed images is affected by the PSF corresponding to the modulating apertures.

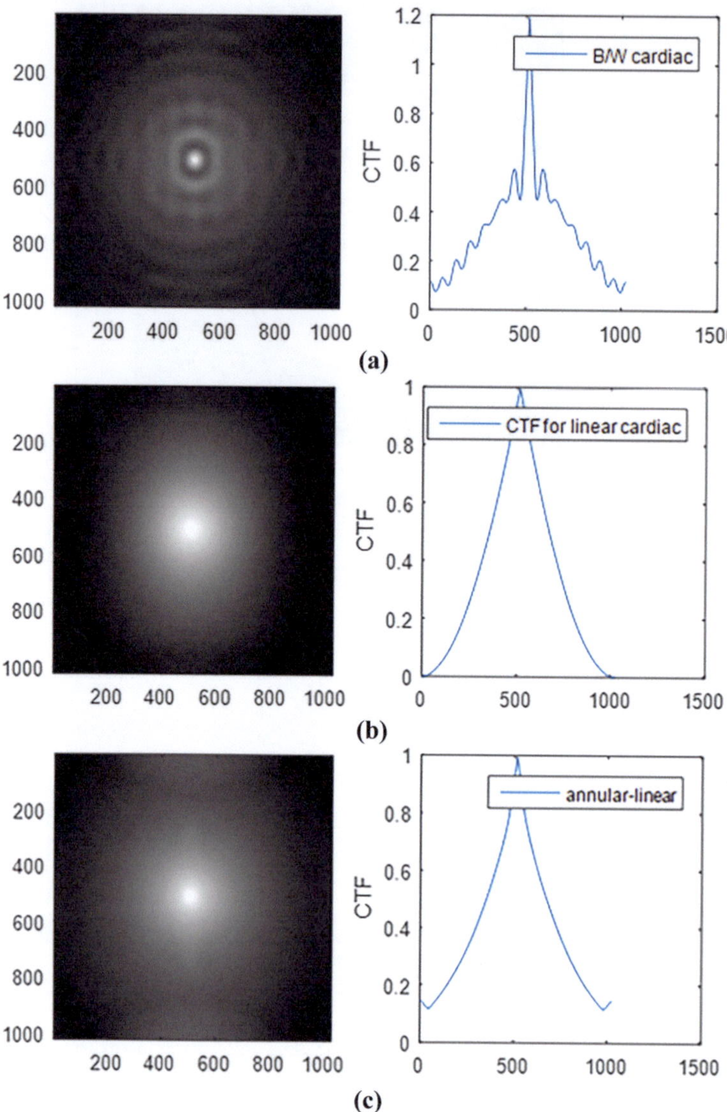

Fig. 7.6 The normalized CTF image and its line plot corresponding to the B/W cardiac aperture are shown in Fig. 7.6a **b** The normalized CTF and its line plot corresponding to the linear cardiac aperture are shown in Fig. 7.6b **c** The normalized CTF and its line plot corresponding to the annular-linear cardiac aperture are shown in Fig. 7.6c **d** The normalized CTF and its line plot corresponding to the cardiac annular aperture are shown in Fig. 7.6d **e** The normalized CTF and its line plot corresponds to the transparent cardiac aperture are shown in Fig. 7.6e

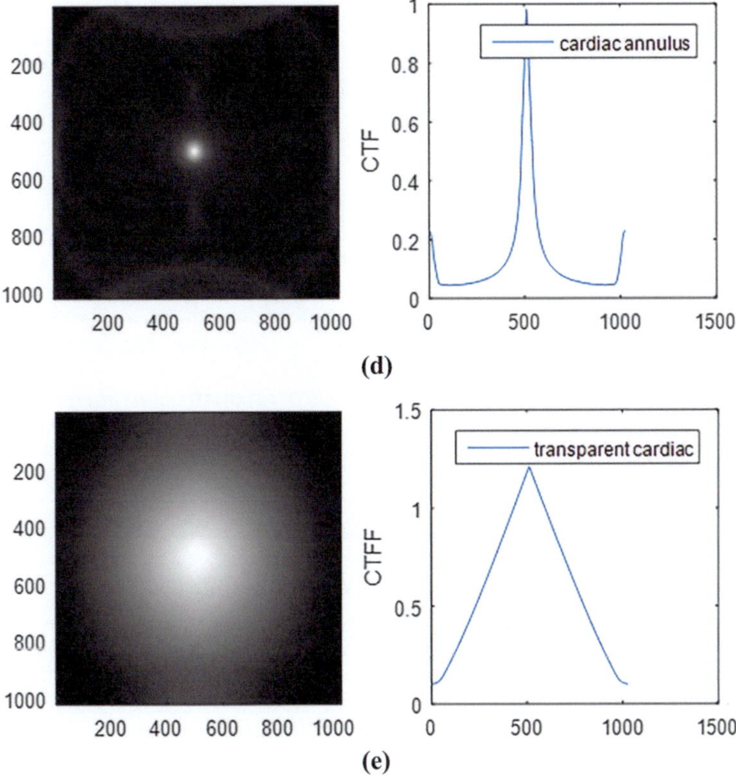

(d)

(e)

Fig. 7.6 (continued)

7.5 Conclusion

First, the PSF is affected by the shape obtained by the modulated apertures.

Second, the CTF in the B/W concentric cardiac annuli has a distinguished fringing shape compared to the other modulating apertures.

Third, the reconstructed images of B/W concentric cardiac annuli, and in the case of linear distribution, have a compromise of contrast and resolution compared to the transparent cardiac apertures.

Finally, the annular cardiac apertures give better-resolved images, while the contrast is poorer when compared with the transparent cardiac aperture, which has better contrast but poorer resolution.

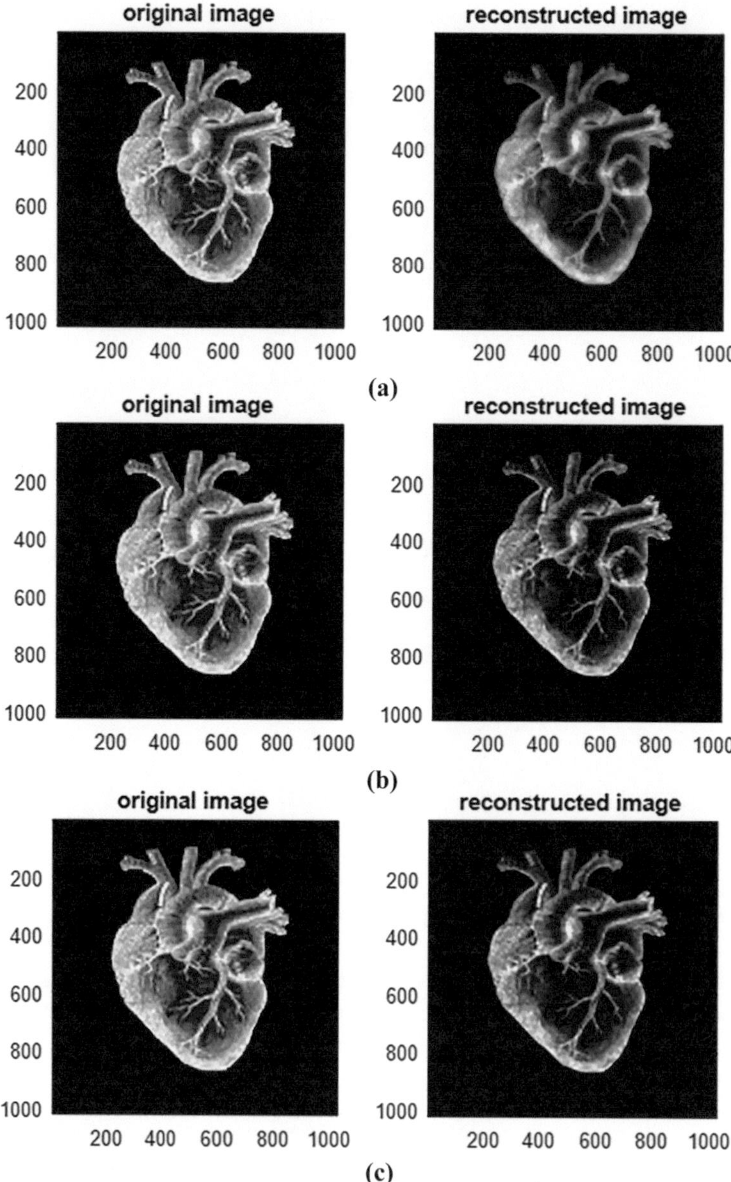

Fig. 7.7 a A reconstruction of the original cardiac image using the B/W cardiac aperture shown in Fig. 7.1a **b** A reconstruction of the original cardiac image using the linear cardiac aperture shown in Fig. 7.2a **c** A reconstructed cardiac image using the aperture composed of a linear distribution surrounded by an annulus shown in Fig. 7.3a **d** A reconstruction of the original cardiac image using the annular cardiac aperture shown in Fig. 7.4a **e** A reconstruction of the original cardiac image using the transparent cardiac aperture shown in Fig. 7.5a

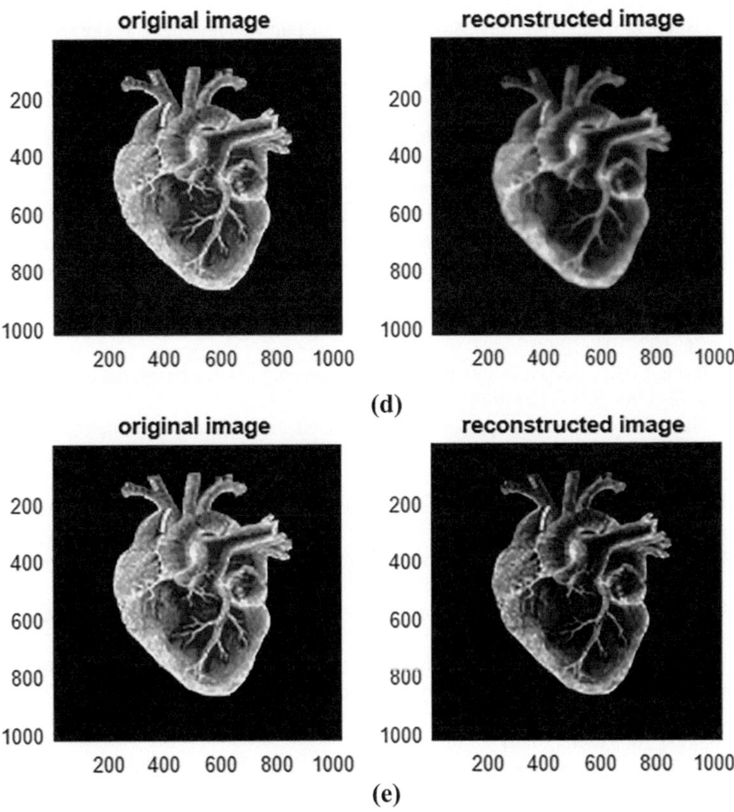

Fig. 7.7 (continued)

References

1. M. Minsky, Microscopy apparatus. US Patent 3,013,467. (1957)
2. C.J.R. Sheppard, A. Choudhury, Image formation in the scanning microscope. Optica Acta: Int. J. Opt. **24**(10), 1051–1073 (1977)
3. C.J.R. Sheppard, Super-resolution in confocal imaging. Optik **80**(2), 53–54 (1988)
4. C.J.R. Sheppard, T. Wilson, Depth of field in the scanning microscope. Opt. Lett. **3**(3), 115–117 (1978)
5. C.J.R. Sheppard, H.J. Matthews, Imaging in high-aperture optical systems. J. Opt. Soc. Am. A **4**(8), 1354–1360 (1987)
6. C.J.R. Sheppard, T. Wilson, Image formation in scanning microscopes with partially coherent source and detector. Optica Acta: Int. J. Opt. **25**(4), 315–325 (1978)
7. C.J.R. Sheppard, M. Gu, Aberration compensation in confocal microscopy. Appl. Opt. **30**(25), 3563–3568 (1991)
8. C.J.R. Sheppard, T. Wilson, Gaussian-beam theory of lenses with annular aperture. IEE J. Microwaves, Opt. and Acoustics. **2**(4), 105–112 (1978)
9. N. Chen, C.H. Wong, C.J.R. Sheppard, Focal modulation microscopy. Opt. Express **16**(23), 18764 (2008)
10. C.J.R. Sheppard, A. Choudhury, Annular pupils, radial polarization, and super-resolution. Appl Opt. **43**(22), 4322 (2004). https://doi.org/10.1364/AO.43.004322

11. C.J.R. Sheppard, Z.S. Hegedus, Axial behavior of pupil-plane filters. J. Opt. Soc. Am. A **5**, 643 (1988)
12. W.T. Welford, Use of annular apertures to increase focal depth. J. Opt. Soc. Am. **50**(8), 749–753 (1960)
13. H.F.A. Tschunko, Imaging performance of annular apertures. Appl. Opt. **13**(8), 1820–1823 (1974)
14. A.M. Hamed, Speckle imaging of annular Hermite Gaussian laser beam. Pram. J. Phys. **95**, 202 (2021). https://doi.org/10.1007/s12043-021-02231-9
15. A.M. Hamed, T.A. Al-Saeed, Image analysis of modified Hamming aperture: application on confocal microscopy and holography. J. Modern Opt. **62**, 801 (2015). https://doi.org/10.1080/09500340.2015.1007102
16. N. Barakat, A.M. Hamed, H. El Ghandour, Study of fluid flow using speckle interferometry. Optik **76**, 102–104 (1987)
17. A.M. Hamed, M.A. Saudy, Computation of surface roughness using optical correlation. Pram. Journal Phys. **68**, 831–842 (2007)
18. A.M. Hamed, Formation of speckle images formed for diffusers illuminated by modulated apertures (circular obstruction). J. Modern Opt. **56**, 1633–1642 (2009). https://doi.org/10.1080/09500340903277792
19. A.M. Hamed, Discrimination between speckle images using diffusers modulated by some deformed apertures: Simulations. Opt. Eng. **50**, 1–7 (2011). https://doi.org/10.1117/1.3530085
20. A.M. Hamed, Speckle imaging using aperture modulation. Publisher Springer Nature, (2024). https://springer.com/book/10.1007/978-3-031-58300-1
21. A.M. Hamed, J.J. Clair, Image and super-resolution in optical coherent microscopes. Optik **64**, 277–284 (1983)
22. A.M. Hamed, J.J. Clair, Studies on optical properties of confocal scanning optical microscope using pupils with radially transmission distribution. Optik **65**, 209–218 (1983)
23. A.M. Hamed, Resolution and contrast in confocal optical scanning microscopes. Opt. Laser Technology **16**, 93–96 (1984)
24. A.M. Hamed, Improvement of point spread function (PSF) using linear quadratic aperture. Optik **131**, 838–849 (2017). https://doi.org/10.1016/j.ijleo.2016.11.201
25. N.C. Pégard et al., Compressive light-field microscopy for 3D neural activity recording. Optica **3**, 517 (2016)
26. T. Seidel, J.C. Edelman, and F.B. Sachse, Analyzing remodeling of cardiac tissue: A comprehensive approach based on confocal microscopy and 3D reconstructions. Ann. Biomed. Eng., **44**(5), 1436–1448 (2016). https://doi.org/10.1007/s.10439-015-1465-6
27. C. Huang, S. Wasmund et al., Case report, Catheterized fiber-optics confocal microscopy of the beating heart in situ. Cardiovasc. Imaging **10**, 1225–1236 (2017)
28. C.G. Refael, R.E. Woods, Digital image processing, 2nd edn, p. 07458. Prentice Hall, Upper Saddle River, New Jersey (2001)
29. A.M. Hamed, Investigation of a concentric black and white hexagonal pupil and its application in confocal microscopic imaging. Opt. and Quant. Electronics **55**, 1279 (2023)

MIX
Papier aus verantwortungsvollen Quellen
Paper from responsible sources
FSC® C105338

If you have any concerns about our products,
you can contact us on
ProductSafety@springernature.com

In case Publisher is established outside the EU,
the EU authorized representative is:
Springer Nature Customer Service Center GmbH
Europaplatz 3, 69115 Heidelberg, Germany

Printed by Libri Plureos GmbH
in Hamburg, Germany